Chemical Reactions

Following the course of a reaction

As a chemical reaction proceeds, the concentration of products increases and the concentration of reactants decreases. This can be shown graphically.

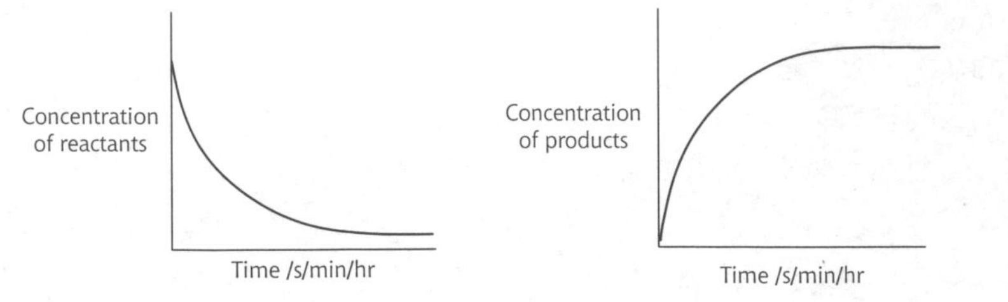

The rate of a chemical reaction can be followed by measuring changes in **concentration, mass** or **volumes of reactants or products**. For example, carbon dioxide gas is produced when calcium carbonate reacts with hydrochloric acid. The reaction can be followed by measuring either the loss in mass (figure1) due to the evolution of carbon dioxide, or the volume of carbon dioxide produced (figure 2).

$CaCO_3(s) + 2HCl(aq) \longrightarrow CaCl_2(aq) + H_2O(l) + CO_2(g)$

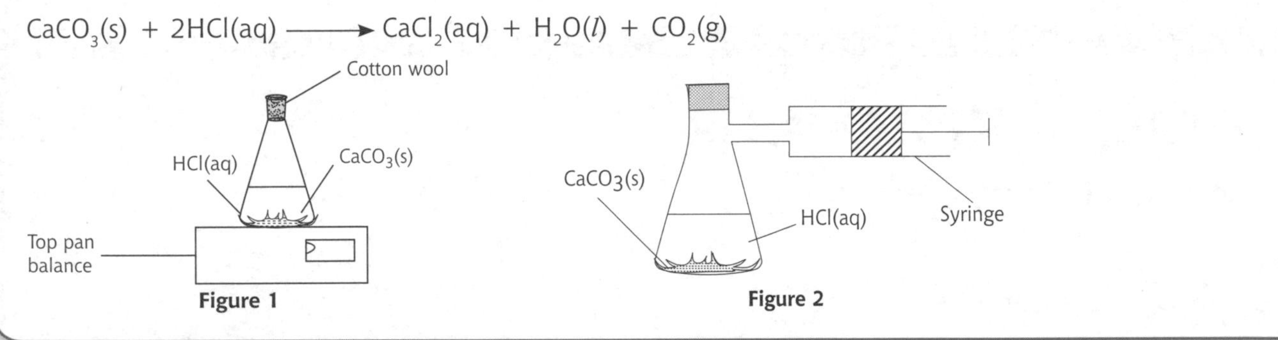

Figure 1 **Figure 2**

Calculating reaction rates

Calculating average rates

$$\text{Average rate} = \frac{\text{change in products or reactant (mass/volume/concentration)}}{\text{time taken for change}}$$

The graph on the right shows the volume of gas given off during a chemical reaction.

Calculate the average rate of the reaction over the first 10 seconds.

$$\text{Rate} = \frac{1.5\ cm^3}{10\ s} = 0.15\ cm^3\ s^{-1}$$

What is the average rate of the reaction between 20 and 30 seconds?

$$\text{Rate} = \frac{2.5 - 2.0\ cm^3}{30 - 20\ s} = \frac{0.5\ cm^3}{10} = 0.05\ cm^3\ s^{-1}$$

The rate of a reaction is greatest at the **beginning** due to the **high concentration** of reactants.

Relative rate

Some reaction rates are expressed as the reciprocal of time: 1/time (s^{-1}, min^{-1} or hr^{-1}), i.e. the relative rate. This occurs when the reaction is stopped at the same point, as in PPA 1 and PPA 2.

For example, a relative rate for a reaction at 20°C was 20 s^{-1}.

The Top Tip shows how to calculate the time taken for this reaction to take place.

Factors affecting rate 1

The rate of a reaction can be altered by changing the concentration or particle size of one or more of the reactants. Temperature can also alter the rate of a reaction.

Collision theory

Before a chemical reaction can take place the reactant molecules must collide successfully. Take the reaction of methane with oxygen:

$$CH_4(g) + 2O_2(g) \longrightarrow CO_2(g) + 2H_2O(l)$$

When you turn on the bunsen burner, methane molecules are released and collide with oxygen molecules in the air. The molecules collide but do not react until you release a spark from a bunsen lighter.

Energy is needed to break the bonds in the colliding particles. When they react, new bonds are made and energy is released. Colliding reactant particles must collide with a certain minimum kinetic energy that will break the bonds in the reactants. This minimum energy is known as the **activation energy**, $\mathbf{E_a}$. If the reactant particles do not possess this energy then they will collide but not react, resulting in an unsuccessful collision. Energy can be supplied by a spark or heat, resulting in the reactant particles having sufficient energy to overcome this activation barrier.

Collision theory and surface area

Chemical reactions occur on the surface of the reactants. As the particle size of a reactant decreases, the surface area increases. The increase in surface area results in an increase in collisions and therefore the reaction rate increases, e.g.:

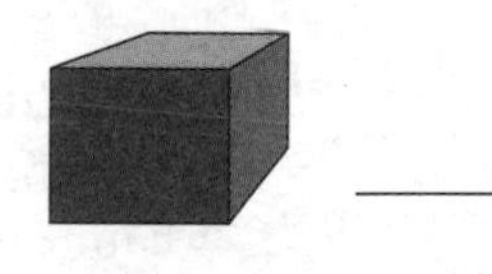

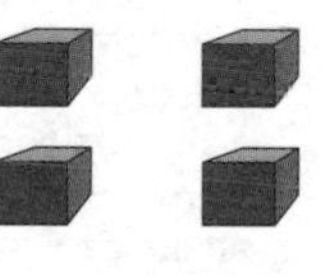

One cube has 6 surfaces where collisions can take place.

Quartering the cube results in 24 surfaces where collisions can take place.

Collision theory and concentration

If you increase the concentration of one or more reactants, you increase the number of particles in a fixed volume. More particles collide and there is an increase in the reaction rate. The rate at the beginning of the reaction is greatest due to the high concentration of reactants. As reactants are converted to products, the concentration of reactants decreases, resulting in a decrease in the number of successful collisions.

Questions

1. At which point during a reaction will the rate be the quickest? Explain.
2. The graph shows the mass of CO_2 produced when Na_2CO_3 reacts with HCl.
   (a) Calculate the average rate between 10 and 20 seconds.
   (b) At what time did the reaction stop?

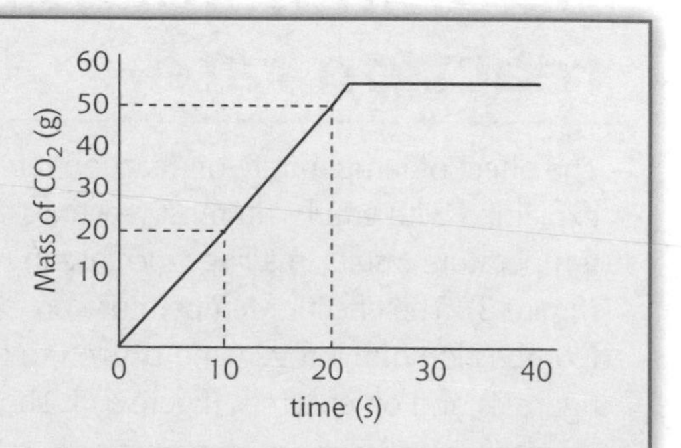

Factors affecting rate 2

Temperature

Temperature is a measure of the average kinetic energy of the particles of a substance.

The kinetic energy of molecules in a reaction mixture varies greatly. Most of the particles will have energy values near the average, whilst a small percentage will have values significantly more or less than the average.

Energy distribution graphs

The effect of temperature on the kinetic energy of the reacting molecules can be explained by energy distribution diagrams. The shaded area in the energy diagram on the right shows the particles with sufficient energy to overcome the activation barrier, E_a, when they collide they will do so successfully.

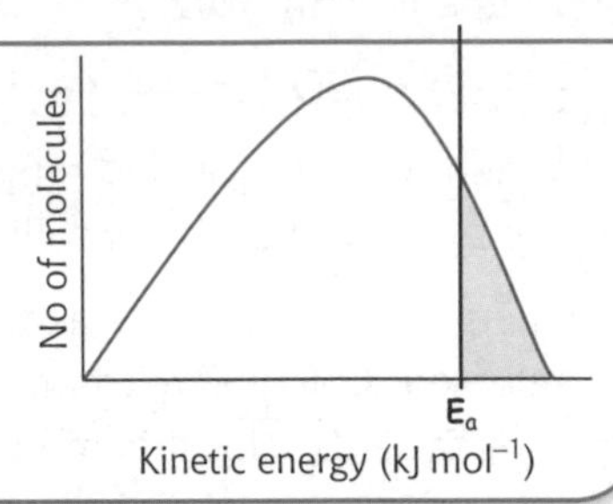

Effect of changing temperature

The energy distribution diagram below shows the effect of varying temperature

At lower temperatures the kinetic energy of the particles decreases. The number of particles with energy in excess of the activation energy decreases, resulting in a reduction in successful collisions.

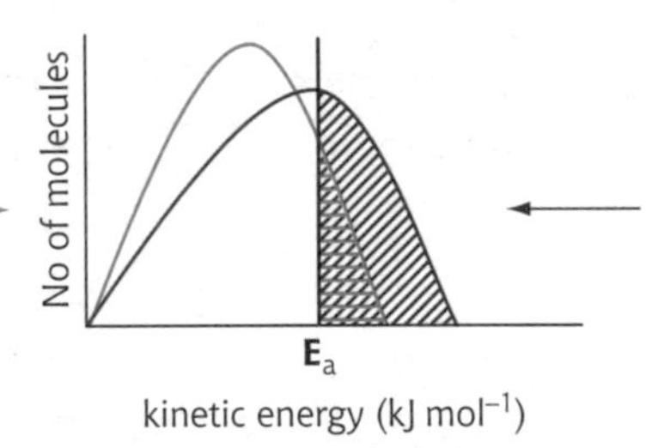

At higher temperatures the kinetic energy of the particles increases. The number of particles with energy in excess of the activation energy increases, resulting in an increase in successful collisions.

Lower temperatures: move the curve to the left and make the curve higher.
Higher temperatures: move the curve to the right and make it flatter than the original.

What is a photochemical reaction?

In some chemical reactions, light can be used to increase the number of particles with energy greater than the activation energy. Photosynthesis occurs when light energy, absorbed by chlorophyll, converts carbon dioxide and water to oxygen and glucose. In Standard Grade or Intermediate 2 you studied the reaction between alkanes and bromine water, which requires light energy. The gases hydrogen and chlorine explode when exposed to high-energy light.

Diagrammatic representation of temperature on reaction rate

The effect of temperature on reaction rate can be explained with graphs. In most reactions a small rise in temperature results in a large increase in reaction rate (figure 3). The effect of temperature on explosive reactions (combustion of hydrogen and oxygen to form water: figure 4) and on enzymes (figure 5) is shown opposite.

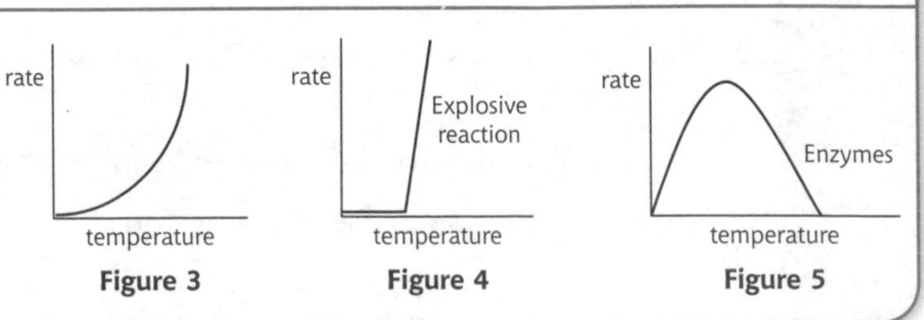

Figure 3 **Figure 4** **Figure 5**

Interpreting rate graphs

Graph 1 in figure 6 below shows the volume of hydrogen released when 50 cm^3 of a 0.2 mol l^{-1} solution of HCl is reacted with excess magnesium granules:

$$Mg(s) + 2HCl(aq) \longrightarrow MgCl_2(aq) + H_2O(l)$$

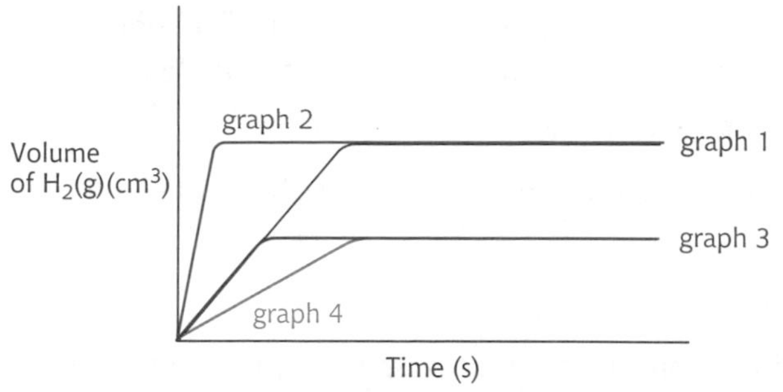

Figure 6

The table below summarises the graphs above.

Graph	Description of graph compared to graph 1		Factors which could bring about the difference in graph from graph 1	
	Rate	**Volume of $H_2(g)$**	**Rate**	**Volume of $H_2(g)$**
2	Faster	Same volume	More concentrated acid. Higher temperature. Magnesium powder.	No change in volume or mass of reactants.
3	Same	Half the volume	No change in concentration, temperature or particle size of Mg.	Half the volume/mass of reactants. 25 cm^3 of 0.2 mol l^{-1} HCl.
4	Slower	Half the volume	Less concentrated acid. Lower temperature. Magnesium ribbon.	Half the volume/mass of reactants. 25 cm^3 of 0.2 mol l^{-1} HCl.

Using a more reactive metal would increase the rate of the reaction. Using a less reactive metal would decrease the rate of reaction.

Questions

1. What term is used to describe the minimum energy required before a reaction can occur?
2. An iron gauze is used to catalyse the reaction between N_2 and H_2 in the Haber process. Why is iron used in the form of a gauze?
3. Draw a diagram to show the energy distribution of molecules in a gas.
   (*a*) Add an additional curve which shows the energy distribution of the gas molecules at a lower temperature.
   (*b*) Explain why a decrease in temperature results in a reduction in the rate of the reaction.

PPA 1 – The effects of concentration changes on reaction rate

Introduction

Hydrogen peroxide(H_2O_2) reacts with potassium iodide (KI) to form iodine molecules.

$H_2O_2(aq) + 2H^+(aq) + 2I^-(aq) \longrightarrow 2H_2O(l) + I_2(aq)$ **Reaction One**

Sodium thiosulphate ($Na_2S_2O_3$) and starch solution are also present in the reaction mixture. The iodine molecules formed in reaction one immediately react with the thiosulphate ions and are converted back into iodide ions.

$I_2(aq) + 2S_2O_3^{2-}(aq) \longrightarrow 2I^-(aq) + S_4O_6^{2-}(aq)$ **Reaction Two**

When all the thiosulphate ions have reacted and therefore been used up, iodine is left, which is indicated by the starch present in the mixture turning a blue-black colour.

Aim

The aim of this PPA was to determine the effect of changing the concentration of potassium iodide in its reaction with hydrogen peroxide.

Procedure

Solutions of sulphuric acid, sodium thiosulphate, starch and potassium iodide were measured into a beaker. The stopwatch was started when the hydrogen peroxide solution was added to the beaker and stopped when the blue-black colour appeared. The concentration of the potassium iodide was changed by using different volumes of potassium iodide and water.

Example results

Volume of KI(aq)(cm^3)	Volume of water	Reaction time (s)	Rate 1/t (s^{-1})
25	0	10	0.100
20	5	18	0.055
15	10	30	0.033
10	15	52	0.019
5	20	105	0.009

The time taken for each reaction was recorded and the relative rate calculated. The rate is taken as a reciprocal of time because the reaction was stopped at the same point in each experiment: the amount of thiosulphate present was the same in each experiment.

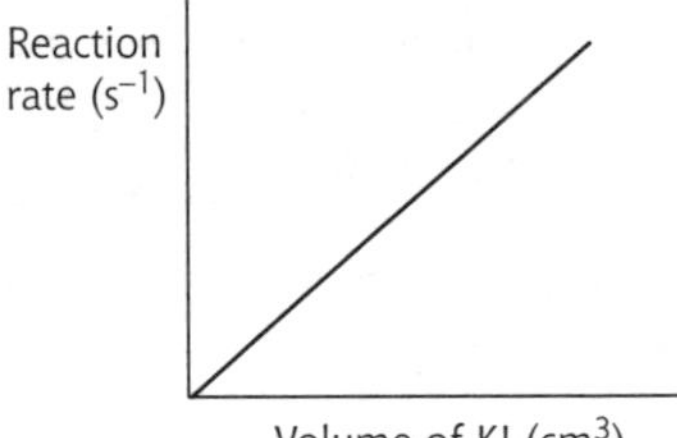

Because the total volume of the reaction mixture did not change, it was assumed that the volume of KI was a measure of its concentration. The results were plotted as a graph and it shows a straight line graph.

For some reactions the rate is directly proportional to the concentration.

Conclusion

As the concentration of potassium iodide increases, the rate of reaction increases.

Evaluation

- The endpoint is easy to determine as the appearance of the blue-black colour is sharp.
- The reaction mixture was placed on a white tile in order to make it easier to detect the colour change.

PPA 2 – The effects of temperature changes on reaction rates

Introduction

Oxalic acid reacts with acidified potassium permanganate solution.

$$5(COOH)_2(aq) + 6H^+(aq) + 2MnO_4^-(aq) \longrightarrow 2Mn^{2+}(aq) + 10CO_2(g) + 8H_2O(l)$$

(purple) (colourless)

The colour changes from purple to colourless as the permanganate ions get used up during the reaction. The number of moles of permanganate ions present is the same in all reactions, so the end point is the time taken for the purple colour to disappear.

Aim

To find out how changing the temperature affects the rate of reaction between oxalic acid and acidified potassium permanganate solution.

Procedure

Volumes of sulphuric acid, potassium permanganate solution and water were measured into a dry beaker. The mixture was warmed to approximately 40°C and placed on a white tile. Oxalic acid was added and a timer immediately started. Timing was stopped when the colour changed from purple to colourless. The temperature at the end of the reaction was recorded. The experiment was repeated at approximately 50°C, 60°C and 70°C.

Example results

Temperature (°C)	Time (s)	Reaction rate (s^{-1})
38	89	0.011
48	31	0.032
61	11	0.090
72	3	0.333

Top Tip

Relative rate = 1/time
Time = 1/relative rate

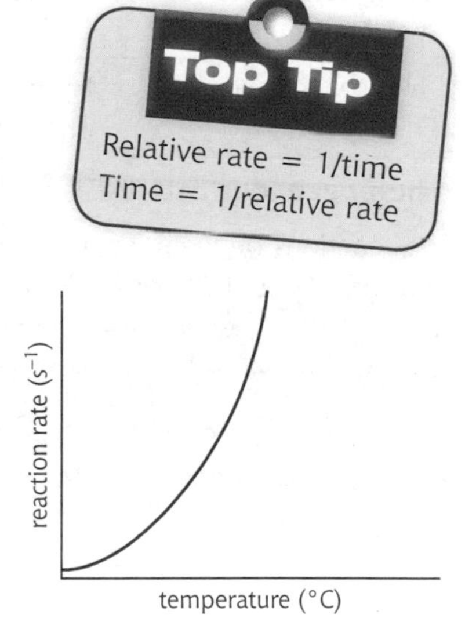

Conclusion

As temperature increases the rate of reaction increases.

A small rise in temperature results in a large increase in the reaction rate.

Evaluation

- The beaker had to be dry to ensure that the concentration of the reactants remained constant.
- At low temperatures the colour change is indistinct. This introduces large timing errors.
- A white tile makes it easier to see the colour change.

Questions

1. The relative rate of a reaction is 0·55 s^{-1}. Calculate the time taken for the reaction to occur.
2. Describe how the concentration of potassium iodide was altered in PPA 1.

Catalysts

A catalyst is a substance which increases the rate of a reaction, whilst remaining chemically unchanged at the end of the reaction.
A **catalyst lowers** the **activation energy** of a reaction, therefore allowing more particles to have sufficient energy to overcome the activation energy barrier. The effect of a catalyst can be shown by the green lines in the energy distribution diagram opposite.

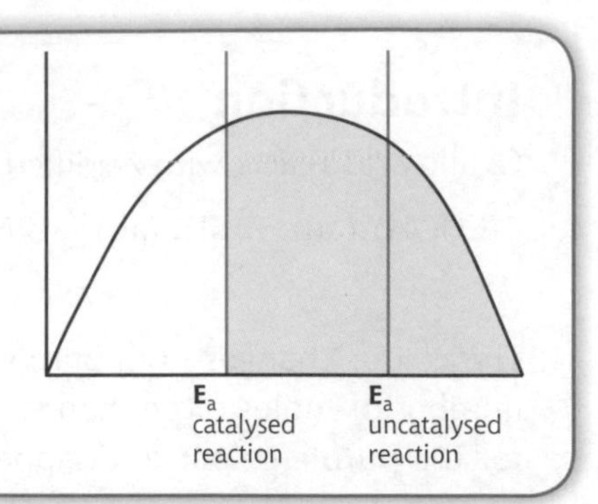

Classifying catalysts

When a catalyst is in a **different state** from the **reactants** it is classified as a **heterogeneous catalyst**. Many catalysts used in industrial processes are examples of heterogeneous catalysts.

Industrial process	Reaction	Catalyst	Catalyst classification
Haber	$N_2(g) + 3H_2(g) \rightleftharpoons NH_3(g)$	$Fe(s)$	Heterogeneous
Ostwald	$4NH_3(g) + 5O_2(g) \rightleftharpoons 4NO(g) + 6H_2O(l)$	$Pt(s)$	Heterogeneous
Catalytic cracking	Breaking down long-chain alkanes to a mixture of short-chain alkanes and alkenes	$Al_2O_3(s)$	Heterogeneous
Contact process	$2SO_2(g) + O_2(g) \rightleftharpoons 2SO_3(g)$	$V_2O_5(s)$	Heterogeneous
Hydrogenation	Hydrogen is bubbled through unsaturated oils to **harden** them.	$Ni(s)$	Heterogeneous

A **homogeneous catalyst** is a catalyst which is in the **same state** as the **reactants**.

The reactions between solutions of potassium sodium tartrate and hydrogen peroxide are catalysed by a solution of cobalt(II) chloride. During the reaction the cobalt(II) ions, which are pink in colour, are oxidised to cobalt(III) ions, which are green in colour. The pink colour of the cobalt(II) ions returns when the reaction has finished, showing that the catalyst has remained unchanged at the end of the reaction. The overall reaction is

$$2H_2O_2(aq) \longrightarrow 2H_2O(l) + O_2(g)$$

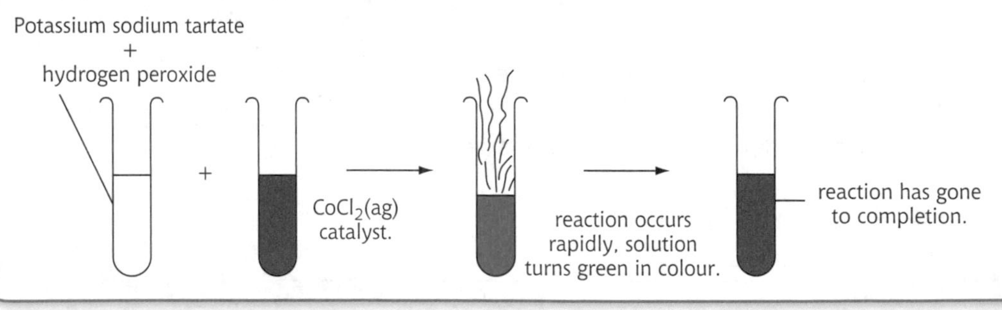

Enzymes

Enzymes are biological catalysts which catalyse many of the reactions which take place in the cells of animals and plants. Enzymes are specific, i.e. each enzyme catalyses only one particular reaction. Enzymes are used in many industrial processes:

- rennin in cheese production
- zymase in the fermentation of glucose to form ethanol
- proteases to tenderise proteins in meat.

Heterogeneous catalysts

It is thought that the following stages occur when reactant molecules adsorb onto the surface of a heterogeneous catalyst. The catalysis takes place at special sites called **active sites** on the catalyst surface. A catalyst provides an **alternative reaction** pathway which results in a **lower activation energy**

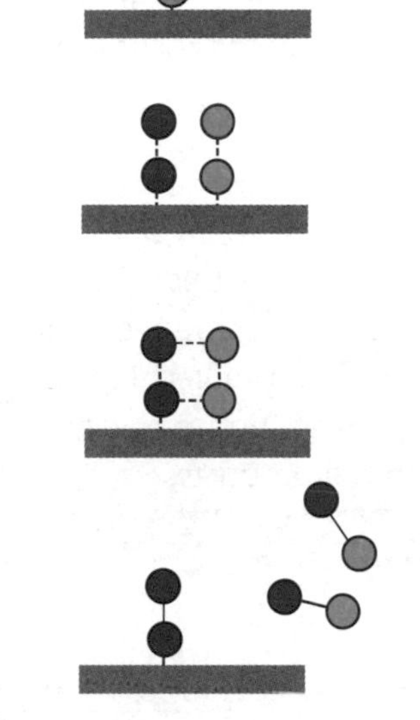

Reactant particles are **adsorbed** onto the surface of the catalyst. The bonds within the reactant particles are weakened.

The reactant molecules are held in the correct collision geometry to allow the reaction to take place.

An activated complex forms.

The products leave the surface of the catalyst leaving it free for more reactant molecules to adsorb onto it.

Catalytic poisoning

If a molecule other than the reactant molecules is **preferentially adsorbed** onto the catalyst's active site, the site is no longer available. This poisoning of the catalyst reduces the activity of the catalyst.

Industrial catalysts can be poisoned by impurities in the reactants. Poisoned ('spent') catalyst can be removed and replaced with fresh catalyst. In some situations the impurities can be removed from the active sites. One way of removing these impurities is to react them with a gas. For example, the catalyst in the catalytic cracking of long-chain alkanes may be poisoned by carbon. The carbon is removed by burning it off in a supply of air. The catalyst is removed, regenerated and then returned to the catalyst chamber.

Catalytic converters are fitted into the exhaust systems of cars to convert poisonous carbon monoxide and oxides of nitrogen into less harmful carbon dioxide and nitrogen. Leaded petrol cannot be used in a car with a catalytic converter because lead adsorbs onto the catalyst surface, blocking the catalyst to the exhaust fumes.

Questions

1. How does a catalyst speed up the rate of a reaction?
2. What is meant by 'poisoning a catalyst'?
3. What is meant by a 'heterogenous catalyst'?

Potential energy diagrams

Enthalpy, Enthalpy changes and activation energy

Every substance contains energy known as enthalpy (H). During a chemical reaction, reactants are used up and products made. Therefore there is a change in enthalpy going from reactant to products. The enthalpy change (shown by the sign ΔH) is the energy difference between products and reactants.

Enthalpy changes

During a chemical reaction, energy is required to break the bonds in the reactant molecules. The energy required to break these bonds is called the **activation energy:** colliding molecules must overcome this activation barrier to form products. When reactant molecules collide, they form a highly energetic arrangement called an **activated complex**. Therefore the activation energy is the energy required by reactant molecules to form an activated complex. This highly energetic complex loses energy and can either go on to form products or fall back into the original reactants. Energy is released when new bonds are formed in the products.

Potential energy diagrams

Potential energy diagrams show the energy pathway for a chemical reaction. The change in enthalpy (ΔH) and activation energy (E_a) for a reaction can be calculated from a potential energy diagram.

Exothermic reactions

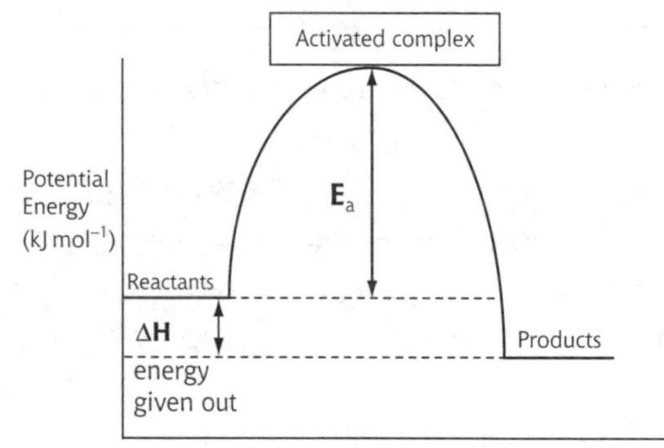

Exothermic changes cause energy, often in the form of heat, to be released into the surroundings.

The enthalpy change (ΔH) is the difference between the enthalpy of the products, H_p, and reactants, H_r.

$\Delta H = H_p - H_r$

The energy of the products is less than that of the reactants. Therefore the enthalpy changes of exothermic reactions are negative, shown by a –ve sign.

Endothermic reactions

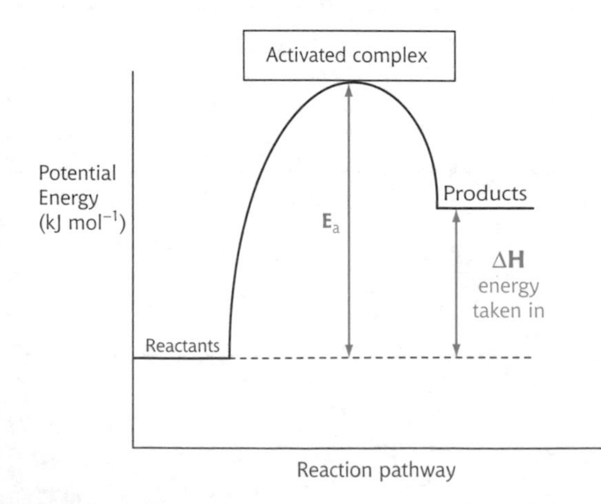

Endothermic changes cause absorption of heat from the surroundings.

The enthalpy change (ΔH) is still the difference between the enthalpy of the products, H_p, and reactants, H_r.

$\Delta H = H_p - H_r$

The energy of the products is greater than that of the reactants. Therefore the enthalpy changes of endothermic reactions are positive, shown by a +ve sign.

Using potential energy diagrams

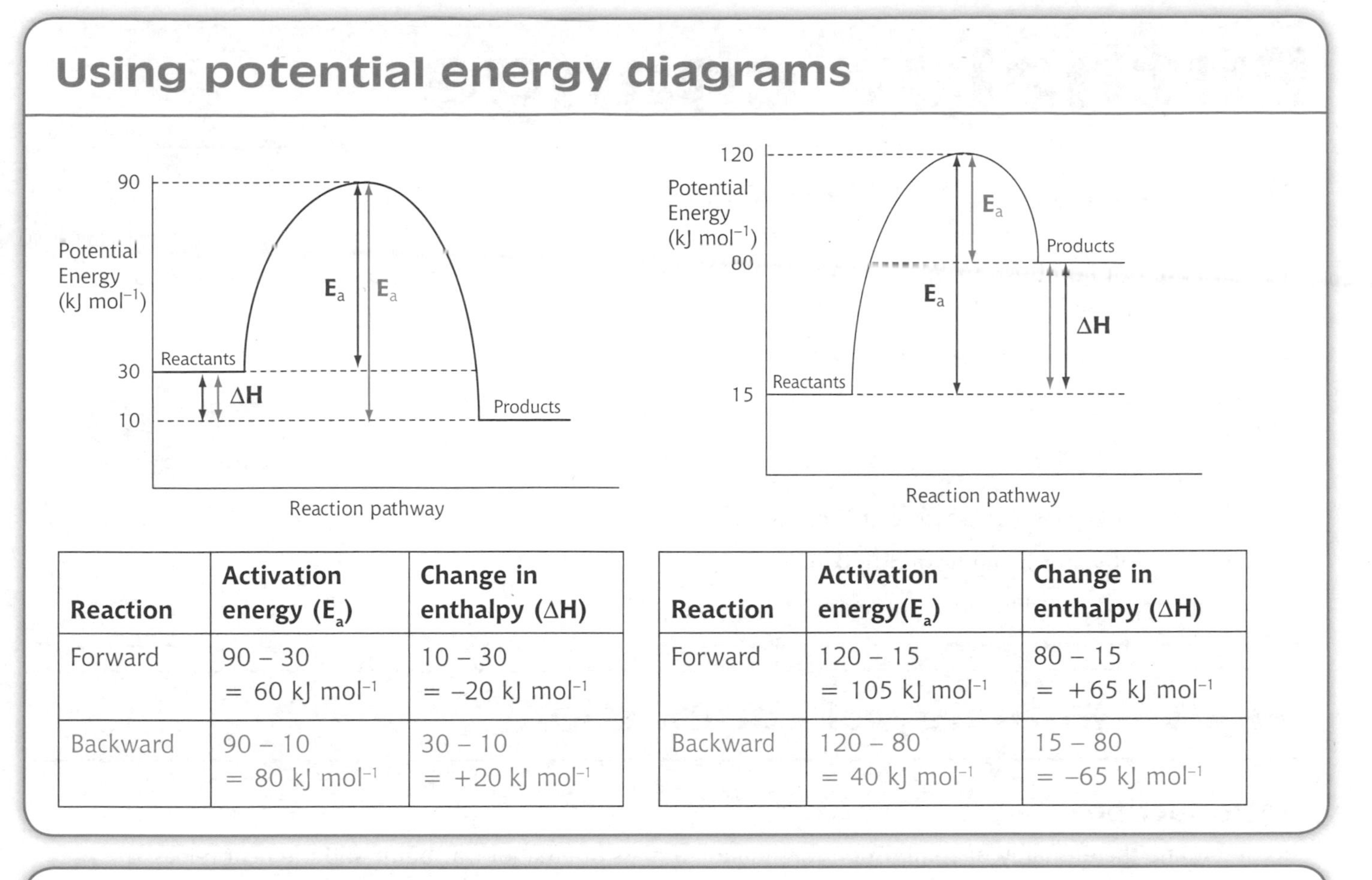

Reaction	Activation energy (E_a)	Change in enthalpy (ΔH)
Forward	90 – 30 = 60 kJ mol^{-1}	10 – 30 = –20 kJ mol^{-1}
Backward	90 – 10 = 80 kJ mol^{-1}	30 – 10 = +20 kJ mol^{-1}

Reaction	Activation energy(E_a)	Change in enthalpy (ΔH)
Forward	120 – 15 = 105 kJ mol^{-1}	80 – 15 = +65 kJ mol^{-1}
Backward	120 – 80 = 40 kJ mol^{-1}	15 – 80 = –65 kJ mol^{-1}

Effect of a catalyst

A catalyst will lower the activation energy (E_a). Therefore more particles have the required energy to overcome the activation barrier and so more product is formed.

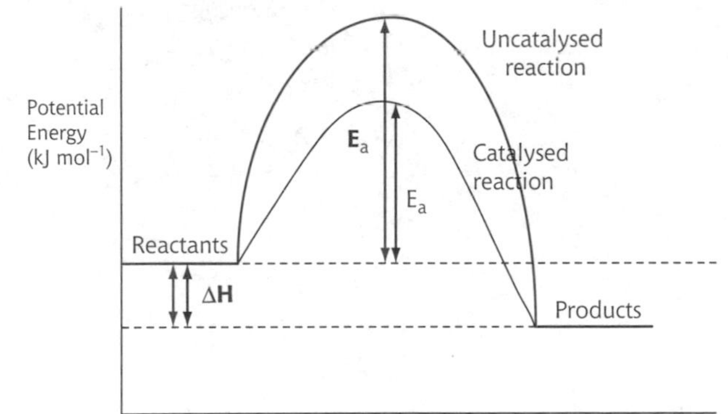

Top Tip

The catalyst reaction pathway must start at reactants and finish at products.

A catalyst will have no effect on the enthalpy change of the reaction.

Questions

Look at the potential energy (PE) diagram.

1. Is this reaction endothermic or exothermic?
2. Calculate the E_a and ΔH for:
   (*a*) the forward reaction
   (*b*) the backward reaction
3. Redraw the PE diagram showing the effect of a catalyst.

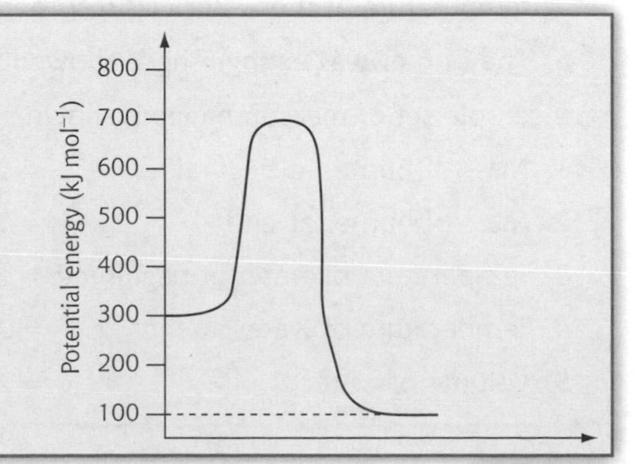

Enthalpy change

Enthalpy of combustion

The **enthalpy of combustion** of a substance is the enthalpy change when **one mole** of a **substance** burns completely in oxygen. The standard enthalpy equation for the combustion of ethanol is:

$C_2H_5OH(g) + 3O_2(g) \longrightarrow 2CO_2(g) + 3H_2O(g)$

The energy produced by a known mass of fuel burning can be calculated using the formula

$E_h = c \times m \times \Delta T$

c = specific heat capacity of water, which has the value 4.18 kJ kg^{-1} $°C^{-1}$

m = the mass of water being heated (in kg)

ΔH = the temperature difference, in degrees centigrade (°C)

Top Tip

$1\ cm^3 = 1\ g$

$1\ g = \frac{1}{1000}\ kg$

PPA 3 – Enthalpy of combustion

Introduction

To determine the enthalpy of combustion for ethanol, a mass of ethanol was burnt and the heat released transferred to a known volume of water. A metal can was used to enhance heat conduction.

Procedure

The apparatus shown was set up.

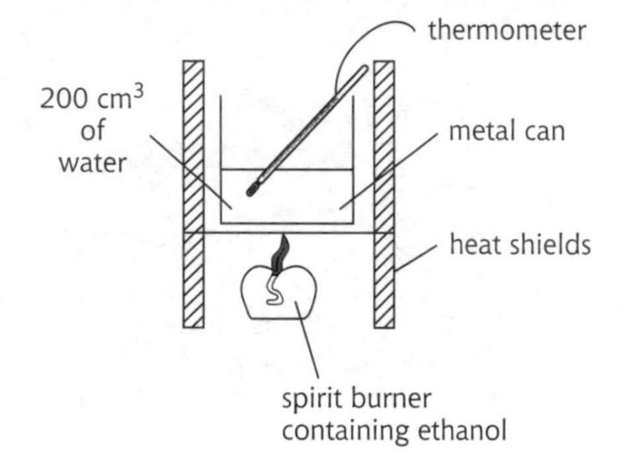

1. The ethanol burner was weighed.
2. A volume of water was measured using a measuring cylinder.
3. The temperature of the water was then measured.
4. The burner was lit and the water heated until a temperature rise of around 10°C was measured. The temperature of the water was recorded.
5. The burner was extinguished, allowed to cool and then reweighed.

A sample set of measurements is shown below:

1. Mass of burner at beginning = 58.50 g
2. Mass of burner at end = 58.19 g
3. Temperature of water at beginning = 18°C
4. Temperature of water at end = 28°C
5. Volume of water. = 200 cm^3

Conclusion

Step 1: Calculate the mass of fuel burnt and temperature change from experimental results.	Temperature change (ΔT) = 10°C Mass of ethanol burnt = 0.31 g
Step 2: Calculate the energy released by known mass (0.31 g) of ethanol (fuel).	$E_h = c \times m \times \Delta T$ $= 4.18 \times 0.2 \times 10$ $= 8.36$ kJ
Step 3: Calculate the mass of one mole of ethanol (fuel).	Molar mass of ethanol $C_2H_5OH = (2 \times 12) + (6 \times 1) + 16 = 46$ g
Step 4: Calculate the enthalpy of combustion – energy released per mole of fuel.	0.31 g of ethanol ⟶ 8.36 kJ 46 g of ethanol (1 mole) ⟶ $\frac{46 \times 8.36}{0.31} = -1{,}240.5$ kJ mol^{-1}

Top Tip

–ve sign if reaction is exothermic.

+ve sign if reaction is endothermic.

Evaluation

The standard enthalpy of combustion for ethanol (given in the data booklet) is –1367 kJ mol^{-1}. The value obtained experimentally is much lower in value due to heat loss to the surroundings and/or incomplete combustion. Ethanol is flammable and should not be placed near a Bunsen burner.

Enthalpy of solution

The enthalpy of solution of a substance is the enthalpy change when one mole of the substance dissolves in water. The enthalpy of solution for potassium hydroxide is shown below. The equation for the reaction is

$KOH(s) \longrightarrow K^+(aq) + OH^-(aq)$

Example

5 g of potassium hydroxide was dissolved in 100 cm^3 of water at 19°C. The highest temperature recorded was 31°C. Calculate the enthalpy of solution of potassium hydroxide.

Step 1: Calculate the energy released by known mass (5 g) of potassium hydroxide.	$E_h = c \times m \times \Delta T$ $= 4.18 \times 0.1 \times 12$°C $= 5.01$ kJ
Step 2: Calculate the mass of one mole of potassium hydroxide.	Molar mass of KOH = 39.1 + 16 + 1 = 56.1 g
Step 3: Calculate the enthalpy of solution – energy change released per mole.	5 g of KOH ⟶ 5.01 kJ 56.1 g of KOH (one mole) ⟶ $\frac{56.1 \times 5.01}{5} = -56.3$ kJ mol^{-1}

Calculating enthalpy change

Enthalpy of neutralisation

The enthalpy of neutralisation is the energy change when an acid is **neutralised** to form **one mole** of water. For example, a neutralisation reaction for hydrochloric acid is:

$HCl(aq) + NaOH(aq) \longrightarrow NaCl(aq) + \mathbf{H_2O(\mathit{l})}$

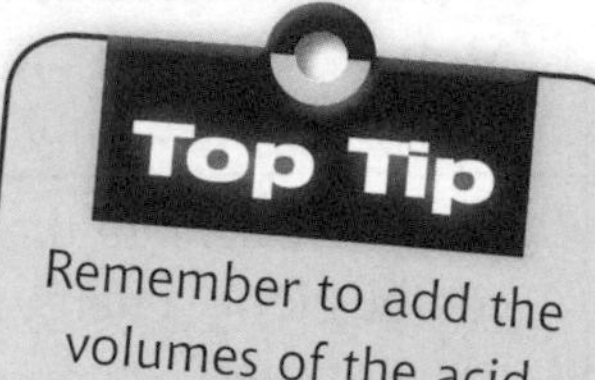

Example

When 100 cm^3 of 1 mol l^{-1} HCl was neutralised by 100 cm^3 of 1 mol l^{-1} NaOH, the temperature rise was by 6 °C. Calculate the enthalpy of neutralisation for this reaction.

Step 1: Calculate the volume and mass of water and the temperature change.	Volume of water = volume of acid + volume of alkali = 100 cm^3 + 100 cm^3 = 200 cm^3 This is 0.2 kg. Temperature change = 6 °C.
Step 2: Calculate the energy released.	$E_h = c \times m \times \Delta T$ $= 4.18 \times 0.2 \times 6\,°C$ $= 5.02$ kJ
Step 3: Calculate the number of moles of water formed.	$HCl(aq) + NaOH(aq) \longrightarrow NaCl(aq) + H_2O(l)$ The mole ratio of acid:alkali:water is 1:1 Moles of acid or alkali = moles of water Moles of acid = $C \times V$ = 1×0.1 = 0.1 moles Moles of water = 0.1 moles
Step 4: Calculate the enthalpy of neutralisation – energy change released per mole of water formed.	0.1 moles of water $\longrightarrow$ 5.02 kJ One mole of water $\longrightarrow 1 \times \frac{5.02}{0.1}$ $= -50.2$ kJ mol^{-1}

Writing standard enthalpy equations

You will use these equations when using Hess's law in Unit 3. Fractions can be used in these reactions, if required to balance the equation.

Enthalpy of combustion

When balancing the equation you must ensure that only **one mole of fuel** is in the balanced equation.

e.g. $\mathbf{C_2H_6} + 3\frac{1}{2}O_2 \longrightarrow 2CO_2 + 3H_2O$

Enthalpy of solution

When balancing the equation only **one mole of solute** should be present.

$\mathbf{(NH_4)_2SO_4(s)} \longrightarrow 2NH_4^+(aq) + SO_4^{2-}(aq)$

Enthalpy of neutralisation

When balancing equations ensure that only **one mole of water** is present in the balanced equation, e.g.:

$HCl(aq) + NaOH(aq) \longrightarrow NaCl(aq) + \mathbf{H_2O(\mathit{l})}$

$^1/_2\ H_2SO_4(aq) + NaOH(aq) \longrightarrow {}^1/_2\ Na_2SO_4(aq) + \mathbf{H_2O(\mathit{l})}$

Questions

1. (*a*) Burning 1.2 g of methanol (CH_3OH) raised the temperature of 500 cm^3 of water by 12°C. Calculate the enthalpy of combustion for methanol.

(*b*) The actual enthalpy of combustion of methanol is –726 kJ mol^{-1}.
Give a reason why the experimental value calculated above is different from the actual value.

2. A pupil is to determine the enthalpy of combustion of ethanol. List the measurements the pupil will need to take.

3. When 0.2 g of methane is burnt, it raises the temperature of 200 cm^3 of water from 19°C to 30°C. Calculate the enthalpy of combustion of methane.

4. Write the standard enthalpy equations for:

(*a*) the combustion of one mole of methane

(*b*) when one mole of $Mg(OH)_2$ dissolves

(*c*) when nitric acid is neutralised by sodium hydroxide

5. The enthalpies of combustion increase by about 660 kJ mol^{-1} on going from methane to ethane and ethane to propane. Give a reason for this constant difference in enthalpies.

6. 5 g of sodium hydroxide is dissolved in 100 cm^3 of water. The temperature of the water rises from 17°C to 30°C. Calculate the enthalpy of solution for sodium hydroxide.

7. When 50 cm^3 of a 1 mol l^{-1} solution of nitric acid was added to 50 cm^3 of a 1 mol l^{-1} NaOH solution, the temperature rose by 7°C. Calculate the enthalpy of neutralisation.

The Periodic Table

The periodic table contains all the naturally as well as man-made elements known so far.

The modern periodic table you use is based on the work of Dmitri Mendeleev in 1869, when he arranged the elements in increasing atomic number. Mendeleev also placed elements with similar chemical properties in groups, leaving gaps so as not to destroy the pattern as more elements were discovered and added.

The diagram below shows the charge from the protons in the nuclei of the elements in the first period and first group in the periodic table. The numbers of electrons and electron shells are also shown.

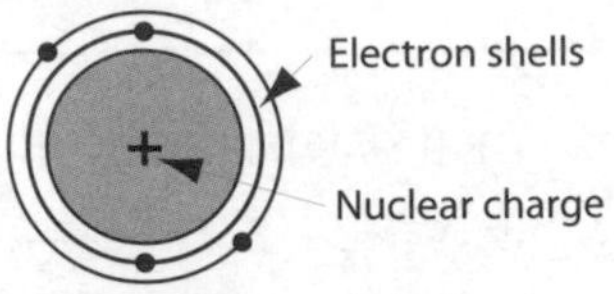

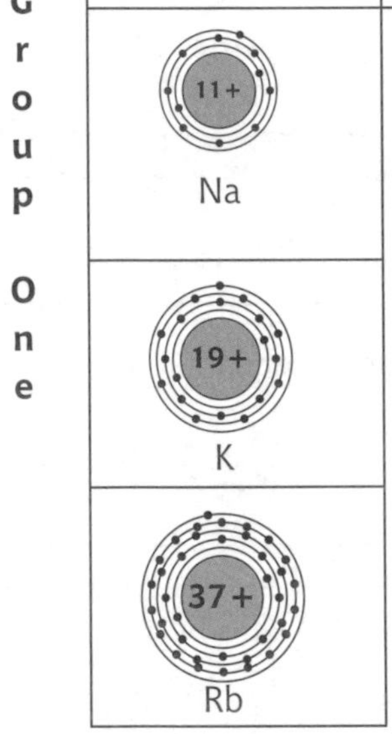

Electron shells
Nuclear charge
+

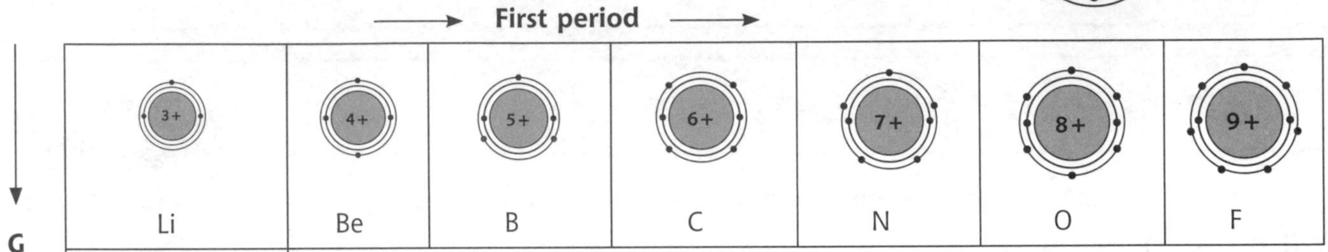

First period

Group One

3+ Li
4+ Be
5+ B
6+ C
7+ N
8+ O
9+ F

11+ Na
19+ K
37+ Rb

Going across a period:

- Nuclear charge increases.
- The same electron shell is filled up.

The attraction between the nucleus and the electrons increases due to the increasing size of the positive nuclear charge.

Going down a group:

- An additional electron shell is added on each time you go down the group.

The outer electrons are further away from the nucleus due to the addition of another outer electron shell. **The attraction of the positive nuclear charge to the outer electrons decreases due to the shielding of the nucleus by these additional electron shells.**

Top Tip

Learn the bold statements.

Atomic size

Across a period: The atomic size **decreases** going across the period. The larger the size of the nuclear charge, the greater the attraction between it and the orbiting electrons, resulting in the electrons being pulled towards the nucleus and the atomic size decreasing.

Down a group: The atomic size **increases** down the group due to the increase in occupied electron shells.

Boiling points, melting points and densities

There are variations in melting point, boiling point and density across periods and down groups.

- Melting and boiling points indicate the strength of bonds between particles. Across a period, melting points increase from group I to group V, after which they drop. Melting and boiling points decrease down group I, due to decreasing attraction between atoms, but melting points increase down group VII as van der Waals attractions (see page 26) between molecules (F_2, Cl_2 and I_2) increase with size.
- Density increases down all groups. It increases across periods until group III, then it decreases.

Ionisation energy

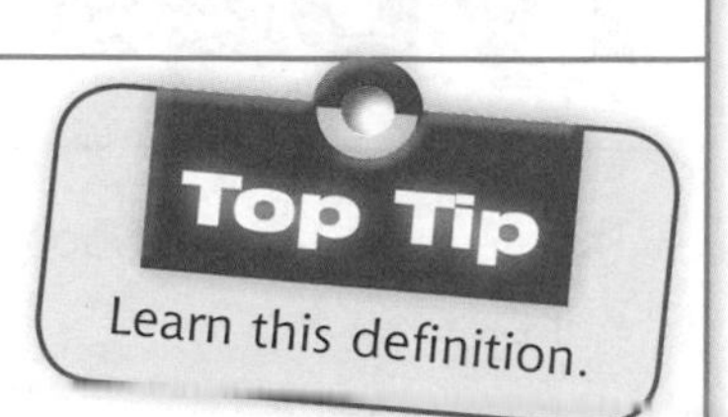

Ionisation energy is the energy required to remove one mole of electrons from one mole of gaseous atoms.

Across the period: The ionisation energy generally increases due to the increasing size of the nuclear charge. The larger the nuclear charge, the greater the attraction for the electrons, resulting in more energy being required to remove the outer electrons.

Down the group: The ionisation energy decreases when going down a group. The attraction of the outer electrons to the nucleus decreases as they move further away from the nucleus. The inner electron shells shield the outer electrons from the attractive effect of the nucleus. Less energy is required to remove the outer electrons.

Calculating ionisation energies

Top Tip

Look at page 10 in the data booklet for ionisation energies.

The first ionisation equation for aluminium is:

$Al(g) \longrightarrow Al^{+}(g) + 1e^{-}$ $\quad \Delta H = +584\,kJ\,mol^{-1}$

The 2nd and 3rd ionisation energies refer to the energies required to move the second and third mole of electrons.

2nd ionisation energy: $Al^{+}(g) \longrightarrow Al^{2+}(g) + 1e^{-}$ $\quad \Delta H = +1830\ kJ\ mol^{-1}$

3rd ionisation energy: $Al^{2+}(g) \longrightarrow Al^{3+}(g) + 1e^{-}$ $\quad \Delta H = +2760\ kJ\ mol^{-1}$

The energy required to turn an aluminium atom into an Al^{3+} ion is the sum of all three ionisation energies ($\Delta H = 584 + 1830 + 2760$).

$Al(g) \longrightarrow Al^{3+}(g) + 3e^{-}$ $\quad \Delta H = +5174\ kJ\ mol^{-1}$

Electronegativity

Atoms of different elements have different attractions for bonding electrons.

Electronegativity is a measure of the attraction an atom involved in a covalent bond has for the shared electrons in that bond.

Across the period: The electronegativity values **increase** across the period.

Down the group: The electronegativity values **decrease** down the group.

Questions

1. Calculate the ionisation for the formation of 1 mole of Mg atoms to Mg^{2+} ions.
2. Crossing the first period, atomic size reduces. Explain why.
3. Explain the trend in ionisation energies down group I.
4. The third ionisation for calcium is 4930 kJ mol l^{-1}. Why is the third ionisation energy so high?

Types of bonding 1

There are 2 classes of bonding:

1. Intramolecular
2. Intermolecular

Intramolecular bonding occurs between atoms and is found inside metallic, ionic and covalent substances. We will look at intermolecular bonding on pages 26–28.

Metallic bonding

As the name suggests, metallic bonding occurs in metals.

Sodium has an electron arrangement of 2,8,1.

The outer electrons can move freely between neighbouring atoms. These electrons are said to be **delocalised.** This leaves behind positive cores.

The metal is held together by the strong forces of attraction between the positive cores and the negatively charged delocalised electrons.

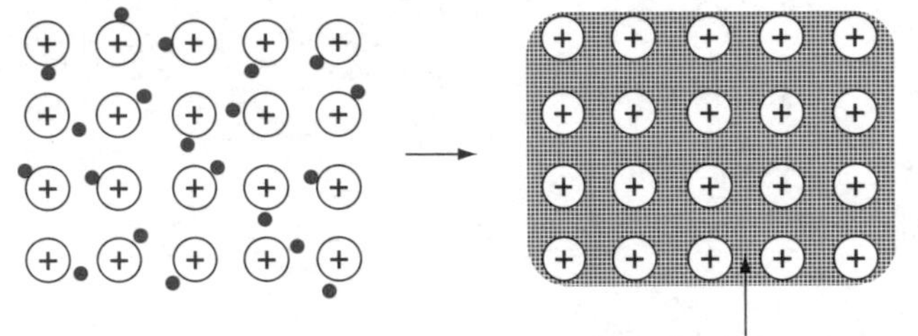

This is sometimes described as 'an array of positive ions in a sea of electrons', but remember: metals are atoms, not ions.

Metallic structure

A metallic structure consists of a giant lattice.

Metals tend to have high melting and boiling points because of the strength of the metallic bond. The strength of the bond varies from metal to metal and depends on the number of electrons that each atom 'delocalises' into the sea of electrons and on how the atoms pack together.

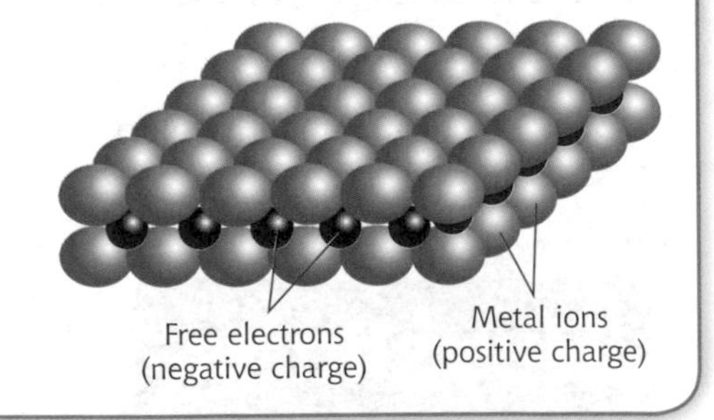

Ionic bonding

This type of bonding occurs between metals and non-metals. Electrons are transferred from the metal to the non-metal in order for both to have full outer electron shells.

Sodium gives its outer electron to chlorine.

Sodium chloride

A sodium atom has an electron arrangement of 2,8,1, and so has **1 outer electron**.

A chlorine atom has an arrangement of 2,8,7, and so has **7 outer electrons.**

When it becomes an ion, sodium loses an electron, so it no longer has equal numbers of electrons and protons. It has one more proton than electron, giving it a charge of +1.

If electrons are lost from an atom, positive ions are formed: Na^+

The chlorine atom has gained an electron, so it now has one more electron than protons. It therefore has a charge of –1.

If electrons are gained by an atom, negative ions are formed: Cl^-

Sodium chloride has the ionic formula Na^+Cl^-.

The ionic bond is the electrostatic force of attraction between the positively charged and negatively charged ions.

When writing ionic formula for groups of ions, don't forget the brackets: $(Al^{3+})_2(O^{2-})_3$.

Ionic structure

An ionic structure consists of a giant lattice of oppositely charged ions.

With sodium chloride, each sodium ion is surrounded by 6 chloride ions and each chloride ion is surrounded by 6 sodium ions.

Na^+: Cl^-
6 : 6

This cancels to a 1:1 ratio, hence the formula Na^+Cl^-.

Ionic compounds have high melting and boiling points, because it requires a great deal of heat energy to overcome the millions upon millions of strong ionic bonds.

Types of bonding 2

Covalent bonding

Covalent bonding usually occurs between non-metals only. Non-metal atoms share electrons in order for both atoms to achieve full outer electron shells.

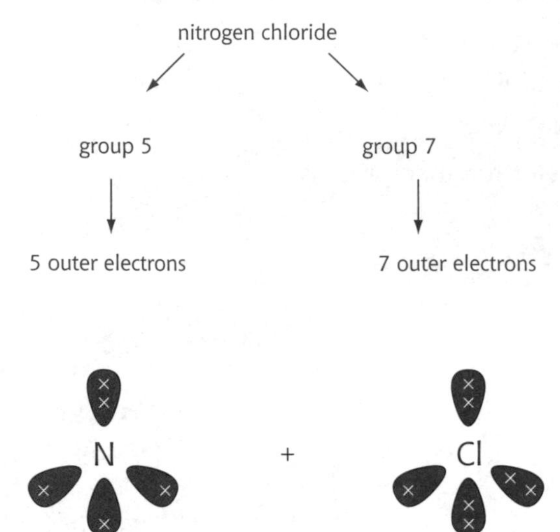

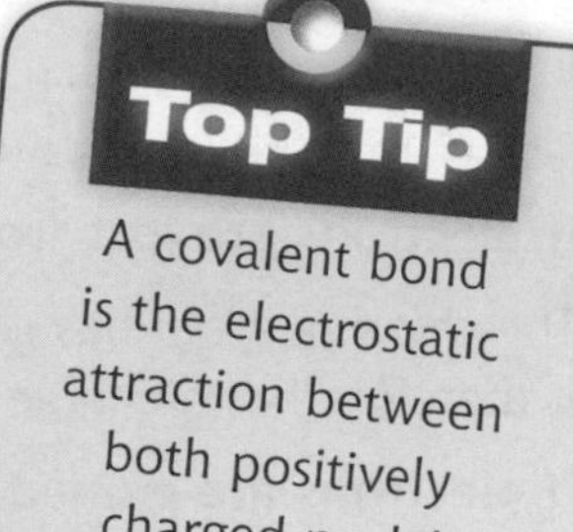

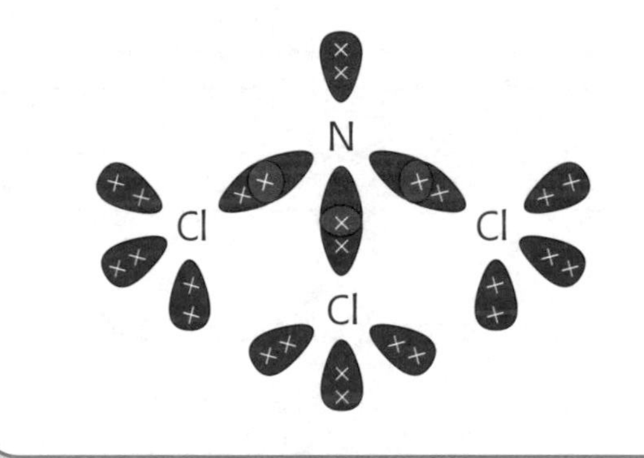

Pure covalent bond

Also called a **non-polar covalent bond.**

In this type of bond, the pair of electrons in the bond is shared equally between the atoms.

Cl —x x— Cl   H —x x— H

To determine if the electrons in the bond are equally shared between the atoms we need to look at the electronegativity values.

The electronegativity value will give a measure of the attraction an atom has for the electrons in the bond.

The higher the value, the more it will pull the electrons towards itself.

In a pure covalent bond, the electronegativity values will be identical.

Polar covalent bond

Look at a molecule of hydrogen fluoride, H——F.

Hydrogen has an electronegativity value of 2·2 and fluorine has an electronegativity value of 4·0. Fluorine has the higher value. This tells us that the fluorine atom will pull the electrons in the bond closer towards it, e.g.

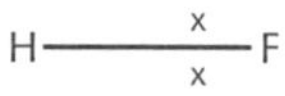

H——$\overset{x}{\underset{x}{}}$—F

This results in the fluorine end of the bond having a slightly negative charge, which makes the hydrogen end of the bond slightly positive.

$\overset{\delta+}{H}$——$\overset{\delta-}{F}$

Top Tip

δ is the symbol delta, and means 'slightly'.

Polar molecules

A polar molecule must contain a permanent polar bond. **However not all molecules which contain polar bonds are polar molecules.**

You need to take into account the **symmetry** of the molecule.

Very simply, **unsymmetrical molecules will be polar**, as in ammonia and water:

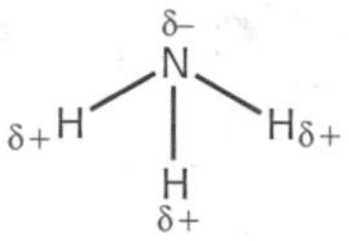

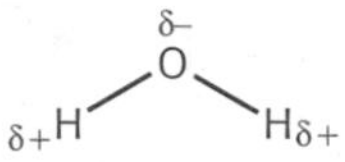

δ– N bonded to δ+ H, δ+ H, δ+ H (ammonia); δ– O bonded to δ+ H, δ+ H (water)

With **symmetrical molecules**, the polarity cancels out and the molecule will be **non-polar**. Carbon dioxide and tetrachloromethane (carbon tetrachloride) are examples of non-polar molecules.

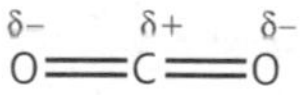

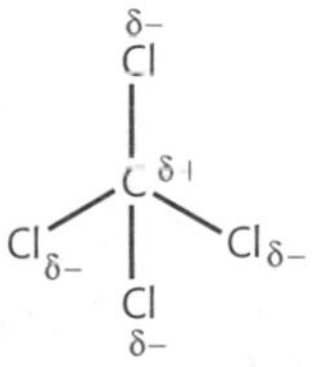

$\overset{\delta-}{O}=\overset{\delta+}{C}=\overset{\delta-}{O}$

δ+ C bonded to four δ– Cl (tetrachloromethane)

Watch out for molecules such as H–C(Cl)(Cl)–Cl (C bonded to H, Cl, Cl, Cl)

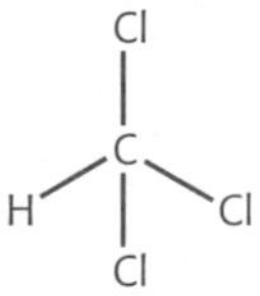

It is easy to miss the hydrogen atom: it is this atom which makes the molecule unsymmetrical and therefore polar.

Questions

1. State what type of covalent bond will exist in the following molecules.

   (a) H——Br (b) O══O (c) C bonded to H, H, H, H

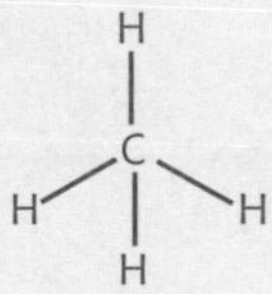

2. State whether the above molecules are polar.

Structure

There are 2 types of covalent structures which can be found in elements and compounds:

1. Network
2. Molecular

The type of structure will affect its properties.

Covalent network elements

This is a giant lattice of covalently bonded atoms.

Carbon, in the form of diamond and graphite, has two covalent network structures: diamond and graphite.

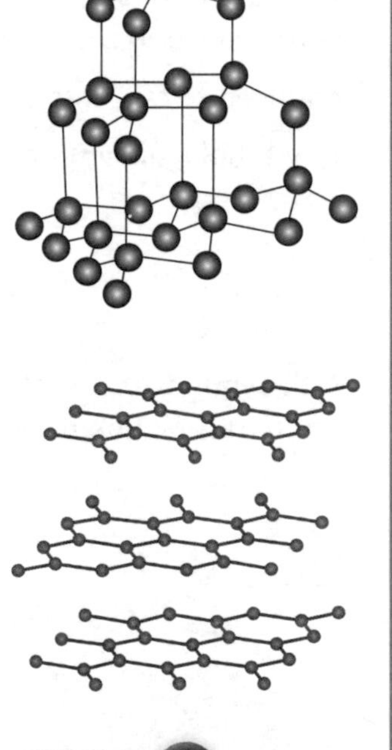

Diamond

- has a very high melting point. Strong carbon–carbon covalent bonds have to be broken throughout the structure before melting occurs.
- is very hard. This is again due to the need to break very strong covalent bonds operating in 3-dimensions.
- doesn't conduct electricity. All the electrons are held tightly between the atoms, and are not free to move.

Graphite

- has a high melting point. In order to melt graphite, you have to break the covalent bonding throughout the whole structure.
- has a soft, slippery feel. It is used in pencils and as a dry lubricant. When you use a pencil, sheets of graphite are rubbed off and stick to the paper.
- conducts electricity. The delocalised electrons are free to move throughout the sheets.

The two other covalent network elements that you have to know are **boron** and **silicon.**

Top Tip

Only three electrons are involved in bonding in graphite. The other electron is delocalised.

Covalent molecular elements

Covalent molecular structures consist of discrete molecules held together by weak intermolecular forces.

These molecules have known numbers of atoms.

The number of atoms present in a molecular substance can vary from:

- very small: H_2, N_2, O_2, F_2

  H—H, N≡N, O=O, F—F

- to a little bit larger – P_4, S_8

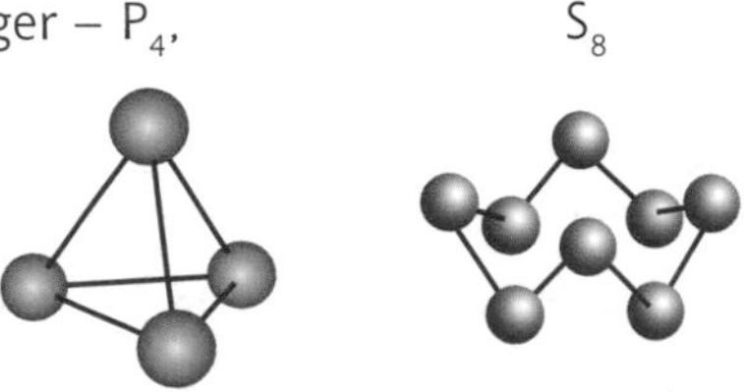

- to very large – fullerenes e.g. C_{60}

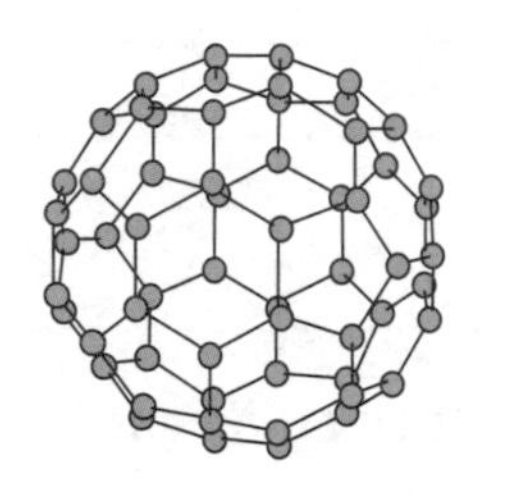

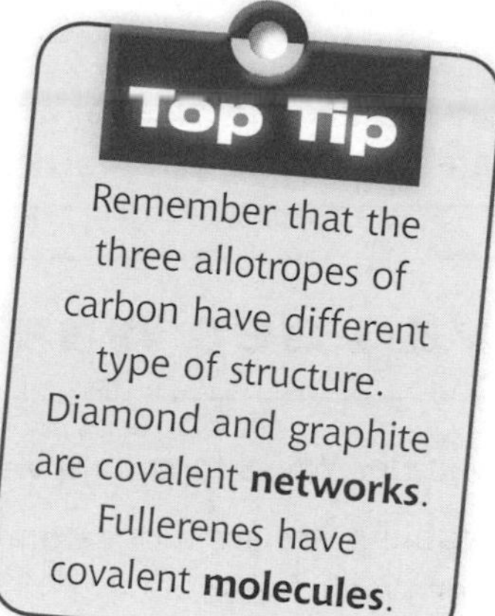

Covalent network compounds

The two main compounds you have to know about are silicon dioxide and silicon carbide.

Silicon dioxide

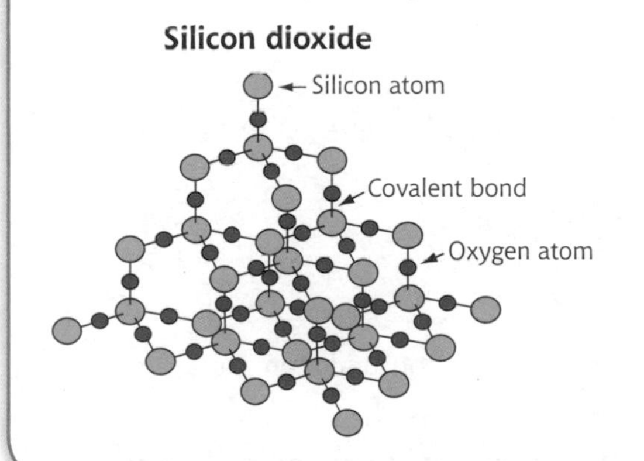

Silicon carbide

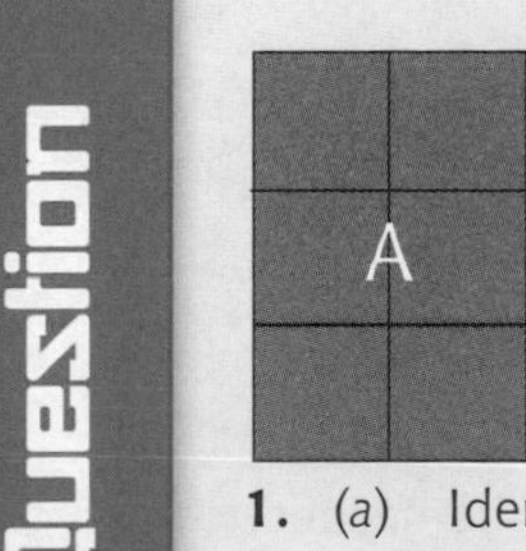

A, D, C, B

1. (*a*) Identify the main types of bonding and structure of the elements found in each block, A, B, C and D. (Clue: look at page 26 for section B.)

   (*b*) Which element in section D can have molecules that do not have the same bonding and structure as the other elements in that block.

Intermolecular forces of attraction

Intermolecular forces of attraction are found between molecules and sometimes between specific atoms. They are weak forces of attraction.

Van der Waals forces

Van der Waals forces are **very weak forces** of attraction that exist between all atoms and molecules.

Noble gases are monoatomic,that is they exist as single atoms and the only type of bonding they have is van der Waals forces.

Neon

An atom of neon has 10 electrons spinning around the nucleus. At any moment in time the electrons will be unequally distributed.

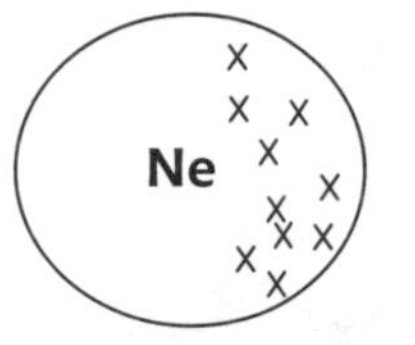

This results in one side of the atom having more electrons than the other, making it slightly negative. As the overall charge on the atom is neutral, then the other side of the atom must be slightly positive.

δ+ δ–

This is called a temporary dipole or an induced dipole and it will affect the orientation of electrons in neighbouring atoms of neon.

δ+ δ– ||||| δ+ δ– ||||| δ+ δ– ||||| δ+ δ– ||| δ+ δ– ||| δ+ δ–

This causes induced dipoles in the other neon atoms, and the weak forces of attraction between the atoms resulting from them are called van der Waals forces.

The strength of the van der Waals forces will affect the melting and boiling points of atoms and molecules.

Van der Waals forces get stronger as the size of the atom or molecule increases.

- **going down a group:** the number of electrons increases, increasing the van der Waals forces

Why does iodine have a higher boiling point than fluorine?

Iodine has more electrons and will therefore have more van der Waals forces and so more energy will be needed to overcome them.

- **in bigger molecules:** more atoms mean more electrons.

Why does ethane (C_2H_4) have a lower melting point than decane ($C_{10}H_{22}$)?

Decane has more atoms and will have more van der Waals forces and so more energy will be needed to overcome them and melt decane.

Permanent dipole–permanent dipole attractions

If a molecule has a permanent dipole, it is said to be a polar molecule.

Hydrogen chloride has a higher boiling than you would expect.

Why?

H—Cl is a polar molecule: chlorine has a much higher electronegativity value than hydrogen and so chlorine will be slightly negative and hydrogen will be slightly positive:

δ+ δ–
H—Cl

Remember to look up the electronegativity values to decide whether a bond is polar or not.

When hydrogen chloride molecules pack together, additional electrostatic forces of attraction appear between the molecules:

δ+ δ– δ+ δ– δ+ δ–
H—Cl H—Cl H—Cl

H—Cl H—Cl
δ+ δ– δ+ δ–

Permanent dipole-permanent dipole attractions are stronger than van der Waals forces. This means more energy is needed to overcome them and we see this in higher melting and boiling points.

Hydrogen bonding

This is the strongest type of intermolecular force of attraction and only occurs between specific atoms. It occurs when a hydrogen atom is bonded to an atom of a strongly electronegative element such as **nitrogen, oxygen** or **fluorine**. These bonds are highly polar due to the large difference in electronegativity between the atoms in the bond

δ– N with δ+H, Hδ+, H δ+ δ– O with δ+H, Hδ+ δ+ δ– H—F

Hydrogen bonds are electrostatic forces of attraction between molecules which contain these highly polar bonds:

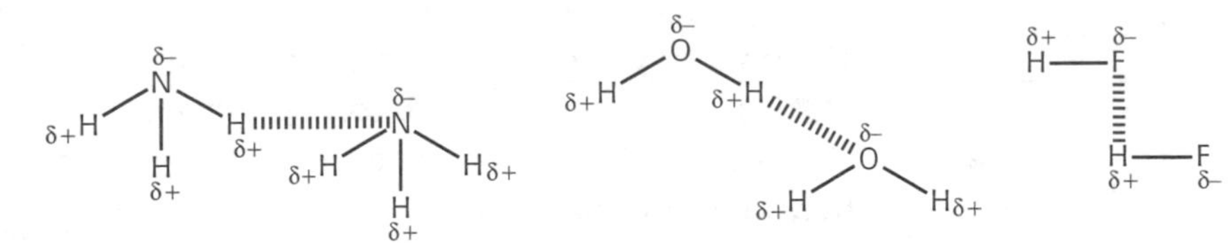

Hydrogen bonds are much stronger than van der Waals forces and stronger than permanent dipole–permanent dipole attractions. This means more energy is needed to boil or melt them. Hydrogen bonds are weaker than covalent bonds.

Properties

The type of bonding present in a substance will determine its properties.

Solubility

Ionic compounds and polar molecules tend to be soluble in polar solvents such as water and insoluble in non-polar solvents such as hexane.

Non-polar substances tend to be soluble in non-polar solvents and insoluble in polar solvents.

Top Tip
Like will dissolve in like.

Melting and boiling points

The melting and boiling point will depend on the type of bonding, or the intermolecular force of attraction, that needs to be overcome.

TYPE OF BONDING	EXAMPLE
bonds 'inside' substance	
covalent network	diamond
ionic	sodium chloride
intermolecular forces of attraction	
hydrogen bonding	water
dipole–dipole attractions	hydrogen chloride
van der Waals forces	neon

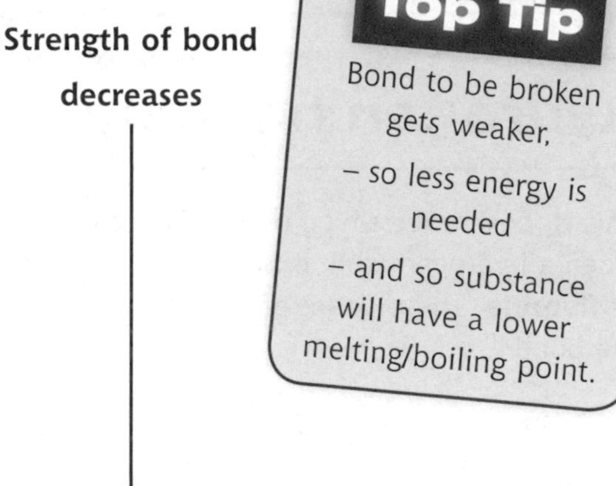

Hydrogen bonding has a considerable effect on the boiling points of water, ammonia and hydrogen fluoride.

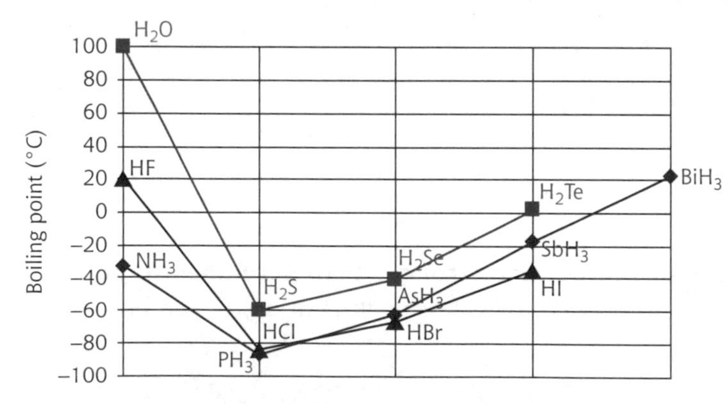

The **boiling point** of the hydride of the first element in each group is **abnormally** high: see H_2O, HF and NH_3. This is due to the presence of **hydrogen bonding**. The intermolecular bonding in the other molecules is the much weaker van der Waals' forces.

Hydrogen bonding's effects can be seen clearly with water. When water freezes, hydrogen bonding causes the water molecules to form an open hexagonal shape.

This results in ice being less dense than water, and it will therefore float on water.

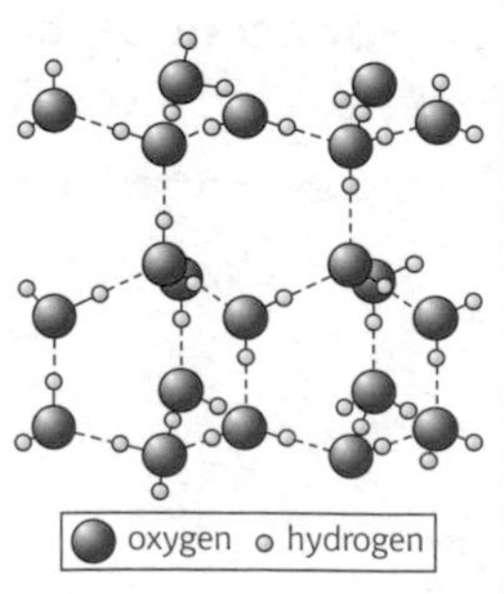

Questions

1. Why can metals conduct electricity?
2. What type of covalent bond exists in
   (*a*) methane CH_4?
   (*b*) tetrafluoromethane, CF_4?
3. Why can ionic compounds, such as molten lead bromide, conduct electricity?
4. What types of structure and bonding exists in
   (*a*) silicon dioxide?
   (*b*) sulphur dioxide?
5. Why does graphite have a much higher melting point than sulphur?
6. What type of bonding exists in solid helium?
7. Why do boiling points increase going down group 7?
8. Why do boiling points increase from methane to hexane?
9. Why does nitrogen hydride have a higher boiling point than phosphorous hydride?
10. Identify the polar molecules:
    (*a*) carbon dioxide
    (*b*) ethanol
    (*c*) tetrafluoromethane
    (*d*) hydrogen iodide
11. What is the bonding and structure of fullerene?
12. Give a use for silicon carbide.
13. Will the following be soluble in water?
    (*a*) sodium fluoride
    (*b*) butane
    (*c*) ammonia
    (*d*) phosphorous hydride
14. Why does ice float?

Reactants in excess 1

Mole revision

The mass of one mole of a substance is its relative formula mass expressed in grams. This is known as its gram formula mass (gfm). The relative formula mass is calculated from the relative atomic masses of the elements in the compound.

Gram formula mass (gfm)

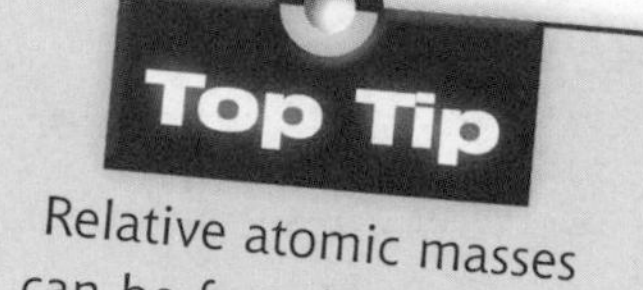

Relative formula mass of $H_2O = (2 \times 1) + 16$

$= 18$

Gram formula mass $= 18$ g

Converting mass to mole and vice versa

In most chemical calculations you will have to convert moles to mass and/or mass to moles.

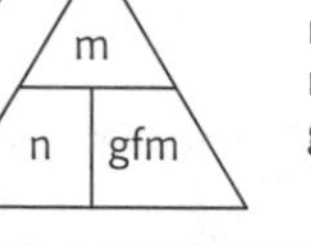

m = mass
n = moles
gfm = gram formula mass

Example

What is the mass of 2 moles of CO_2?	How many moles are present in 250 g of $CaCO_3$?
mass = n × gfm = 2 × 44 = 88 g	moles = mass/gfm = 250/100 = 2.5 moles

Concentration and the mole

A solution is formed when a solute is dissolved in water.

The concentration of a solution can be expressed as mol^{-1} or $g\ l^{-1}$.

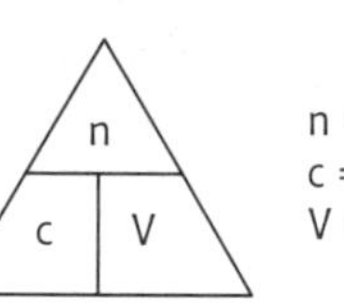

n = moles
c = concentration
V = volume

Examples

1. 40 g of sodium chloride is dissolved in 500 cm³ of water. What is the concentration of the solution?

Step 1: Convert mass to moles.	moles = mass / gfm = 40/58.5 = 0.68
Step 2: Determine concentration in mol l^{-1}.	concentration = mole /volume = 0.68/0.5 = 1.37 mol l^{-1}

2. What mass of sodium carbonate (Na_2CO_3) will form 200 cm³ of a 0.1 mol l^{-1} solution.

Step 1: Determine moles of solute.	Moles = concentration × volume = 0.1 × 0.2 = 0.02 moles
Step 2: Convert moles to mass.	Mass = moles × gfm = 0.02 × 106 = 2.12 g

Calculations from balanced equations

A balanced equation indicates the ratio of moles of reactants and products. The concentration or mass of a product or reactant can be calculated by using the balanced equation.

What is the mole ratio?

If you know the number of moles of one substance, you can determine the moles of another using the balanced equation. This is called the mole ratio.

Examples

1. How many moles of oxygen will react with 0.25 moles of hydrogen?

$2H_2 + O_2 \longrightarrow 2H_2O$	Mole ratio $2H_2 : O_2$ 2 moles : 1 mole 0.25 : 0.125

2. Methane burns in oxygen to produce carbon dioxide gas. Calculate the mass of carbon dioxide produced when 12 g of methane is fully combusted.

$CH_4(g) + 2O_2(g) \longrightarrow CO_2(g) + 2H_2O(l)$

Step 1: Calculate the number of moles of $CH_4(g)$.	n = mass/ gfm = 12/16 = 0.75
Step 2: Identify the mole ratio	CH_4 : CO_2 1 mole : 1 mole 0.75 : 0.75
Step 3: Convert moles of $CO_2(g)$ to mass	m = n × gfm = 0.75 × 44 = 33 g

3. 50 cm^3 of calcium hydroxide solution was neutralised by 25 cm^3 of 1 mol l^{-1} solution of HCl. What is the concentration of the calcium hydroxide solution?

$Ca(OH)_2(aq) + 2HCl(aq) \longrightarrow CaCl_2(aq) + 2H_2O(l)$

Step 1: Calculate the moles of HCl(aq).	n = C × V = 1 × 0.025 = 0.025 moles HCl
Step 2: Identify mole ratio	2HCl : $Ca(OH)_2$ 2 moles : 1 mole 0.025 : 0.0125
Step 3: Calculate the concentration of $Ca(OH)_2(aq)$.	C = n/V = 0.0125/ 0.05 = 0.25 mol l^{-1}

Top Tip

Organise raw data by listing values under the correct reactant or product in a balanced equation before starting your calculation. An alternative is to underline or highlight values in the question.

Reactants in excess 2

Once one or all of the reactants in a chemical reaction are used up, the reaction stops. Any reactant which is left unreacted is said to be in excess.

Reactant in excess calculations

There can be two stages to reactant in excess questions. The stages are:

1. Identifying the reactant in excess.
2. Calculating the mass, volume or concentration of an unknown.

Example

What mass of magnesium chloride is formed when 2 g of magnesium reacts with 100 cm^3 of 1 mol l^{-1} solution of hydrochloric acid?

$Mg(s) + 2HCl(aq) \longrightarrow MgCl_2(aq) + H_2(g)$

Identify the reactant in excess.

Step 1: Determine the numbers of moles of reactants.	Mg Mass = 2 g gfm = 24.3 n = 2/24.3 = 0.08	HCl C = 1 mol l^{-1} V = 0.1 l n = 1×0.1 = 0.1
Step 2: Identify the reactant in excess using mole ratio.	1 mole of Mg ⟶ 2 mole of HCl 0.08 moles of Mg ⟶ 0.16 moles of HCl There are only 0.1 moles of HCl, therefore not all of the magnesium will react. The magnesium is in excess.	

Continue the question using the reactant **not in excess.** Since all of the reactant not in excess **will react,** the number of moles of product will depend on the moles of this reactant.

Step 1: Determine the mole ratio of reactant not in excess and product.	2 HCl : $MgCl_2$ 2 moles : 1 mole
Step 2: Determine moles of product using mole ratio.	0.10 moles : 0.05 moles
Step 3: Calculate the mass of product.	Mass = n × Fm = 0.05×95.3 = 4.756 g

Questions

1. Calculate the gram formula mass of

(*a*) Al_2O_3

(*b*) $(NH_4)_2SO_4$

2. Calculate the number of moles in:

(*a*) 77 g of tetrachloromethane (CCl_4)

(*b*) 110 g of CO_2(g)

3. What is the concentration of the following solutions?

(*a*) a 500 cm^3 solution containing 10 g of NaOH.

(*b*) a 100 cm^3 solution containing 47.6 g of $MgCl_2$.

4. Calculate the mass of NH_4Cl required to make a 200 cm^3 solution with a concentration of 0.1 mol l^{-1}.

5. Methane reacts with oxygen to form carbon dioxide and water. What mass of oxygen is required to produce 11 g of CO_2?

$CH_4(g) + 2O_2(g) \longrightarrow CO_2(g) + H_2O(l)$

6. Magnesium displaces copper from a solution of copper(II) sulphate.

$Mg(s) + CuSO_4(aq) \longrightarrow MgSO_4(aq) + Cu(s)$

Calculate the mass of magnesium required to produce 2.15 g of copper, assuming the copper(II) sulphate solution is in excess.

7. 25 cm^3 of H_2SO_4 was neutralised by 50 cm^3 of a 0.1 mol l^{-1} solution of LiOH. Calculate the concentration of the sulphuric acid.

$2LiOH(aq) + H_2SO_4(aq) \longrightarrow Li_2SO_4(aq) + 2H_2O(l)$

8. Which of the following reactants are in excess?

(*a*) 12.25 g of Mg reacts with 14 g of O_2 to form magnesium oxide.

$Mg(s) + {}^1/_2O_2(g) \longrightarrow MgO(s)$

(*b*) 100 cm^3 of a 1 mol l^{-1} solution of NaOH reacts with 500 cm^3 of a 0.5 mol l^{-1} solution of H_2SO_4.

$2NaOH(aq) + H_2SO_4(aq) \longrightarrow Na_2SO_4(aq) + 2H_2O(l)$

9. Calculate the mass of calcium chloride produced when 10 g of calcium hydroxide is reacted with 50 cm^3 of 1 mol l^{-1} solution of hydrochloric acid.

$Ca(OH)_2(aq) + 2HCl(aq) \longrightarrow CaCl_2(aq) + 2H_2O(l)$

10. Magnesium reacts with sulphuric acid to produce magnesium sulphate and hydrogen.

Calculate the mass of magnesium sulphate produced when 6.1 g of magnesium reacts with 100 cm^3 of 2 mol l^{-1} acid.

$Mg(s) + H_2SO_4(aq) \longrightarrow MgSO_4(aq) + H_2(g)$

11. Calculate the mass of water produced when 3.0 g of ethene burns in 1.6 g of oxygen.

$C_2H_4(g) + 3O_2(g) \longrightarrow 2CO_2(g) + 2H_2O(l)$

The Avogadro constant

In 1 mole of any substance there are 6.02×10^{23} formula units. This number is known as Avogadro's constant and has the symbol L.

What is a formula unit?

To determine what the formula unit is, you need to look at the formula and decide what type of structure it has.

For elements, such as metals and the noble gases, the formula unit is the **atom.**

For covalent molecules, such as nitrogen or decane, the formula unit is the **molecule.**

For ionic compounds, such as sodium chloride, the formula unit is the **ionic formula.**

Substances with the same number of moles will contain the same number of formula units.

Atoms

1 mole of aluminium weighs 27 grams and will contain 6.02×10^{23} atoms

Molecules

1 mole of fullerene, C_{60}, weighs 720 grams and will contain 6.02×10^{23} molecules

Ionic compounds

1 mole of calcium carbonate, $CaCO_3$, weighs 100 g and will contain 6.02×10^{23} $Ca^{2+}CO_3^{2-}$ units

Examples

1. How many atoms are in 2.07 g of lead?

Step 1: Write down the key relationship – lead is a metal and will contain atoms	1 mole of lead contains 6.02×10^{23} atoms
Step 2: Convert moles into mass	207 g of lead contains 6.02×10^{23} atoms
Step 3: Calculate how many atoms of lead are in 2.07 g	2.07 g of lead $= \frac{2.07}{207} \times 6.02 \times 10^{23}$ $= 6.02 \times 10^{21}$ lead atoms

2. How many atoms are in 90 g of water?

Step 1: Write down the key relationship – water, H_2O, is a molecule	1 mole of water contains 6.02×10^{23} molecules
Step 2: Convert moles into mass	18 grams of water contain 6.02×10^{23} molecules
Step 3: Calculate how many molecules are in 90 g	90 grams of water $= \frac{90 \times 6.02 \times 10^{23}}{18}$ $= 3.01 \times 10^{24}$ molecules
Step 4: Calculate how many atoms are in 90 grams. H_2O = 3 atoms	3.01×10^{24} molecules $\times 3 = 9.03 \times 10^{24}$ atoms

3. What mass of water contains 1.806×10^{23} molecules?

Step 1: Write down the key relationship – water is a molecule	6.02×10^{23} molecules are in 1 mole of water
Step 2: Convert moles into mass	6.02×10^{23} molecules are in 18 g of water
Step 3: Calculate the mass	1.806×10^{23} molecules $= \frac{1.806 \times 10^{23}}{6.02 \times 10^{23}} \times 18 = 5.4$ g

4. How many ions are in 37 grams of calcium hydroxide, $Ca^{2+}(OH^-)_2$?

Step 1: Write down the key relationship – $Ca(OH)_2$ contains $1 \times Ca^{2+}$ ion and $2 \times OH^-$ ion	1 mole of $Ca(OH)_2$ contains $3 \times 6.02 \times 10^{23}$ ions $= 1.806 \times 10^{24}$ ions
Step 2: Convert moles into mass	74 grams of $Ca(OH)_2$ contains 1.806×10^{24} ions
Step 3: Calculate the number of ions	37 g of $Ca(OH)_2 = \frac{37}{74} \times 1.806 \times 10^{24}$ ions $= 9.03 \times 10^{23}$ ions

Questions

1. How many molecules are in 0.5 moles of ethanol?
2. How many atoms are in 40 grams of neon?
3. How many ionic units are in 1.06 grams of Na_2CO_3?
4. How many atoms are in 3.2 g of SO_2?
5. What mass of C_4H_8 contains 6.02×10^{24} molecules?
6. What mass of Al_2O_3 contains 1.806×10^{24} ionic units?
7. Which of the following contains the greater number of ions?
   (*a*) 4 g of NaOH
   (*b*) 81.6 g of AlI_3
   (*c*) 19.0 g $MgCl_2$
   (*d*) 40.6 g of Pt_2O

Molar volume

The volume of 1 mole of a gas is known as the **molar volume** and it has the unit, litres mol^{-1}.

The molar volume is the same for all gases at the same temperature and pressure.

For example, 1 mole of chlorine weighs 71 g and will occupy the same volume as 1 mole of helium weighing 4 g at the same temperature and pressure.

Examples

In the following questions, take the molar volume to be 24 litres mol^{-1}.

1. What volume will 280 g of carbon monoxide, CO, occupy?

Step 1: Write down the key relationship – moles to volume	1 mole of CO occupies 24 litres
Step 2: Convert moles into mass	28 g occupies 24 litres
Step 3: Calculate what 280 g is in litres	280 g occupies $\frac{280 \times 24}{28}$ = 240 litres

2. What mass of ethane, C_2H_6, occupies 4.8 litres?

Step 1: Write down the key relationship – volume to moles	24 litres contains 1 mole of ethane
Step 2: Convert moles into mass	24 litres contains 30 grams
Step 3: Calculate what 4.8 litres is in grams	4.8 litres contains $\frac{4.8}{24} \times 30$ = 6 g

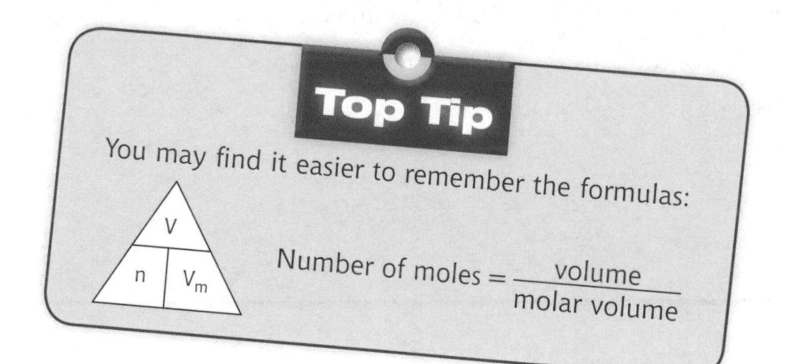

3. What volume will 16 g of helium occupy?

Step 1: Calculate number of moles	$n = \frac{m}{gfm}$ $n = \frac{16}{4} = 4$ moles
Step 2: Convert moles into volume	$V = n \times V_m$ $V = 4 \times 24 = 96$ litres

4. What mass of nitrogen occupies 5 l? (Assume that the molar volume of nitrogen is 24 l mol^{-1}.)

Step 1: Calculate moles of nitrogen	$n = v \div V_m$ $= 5 \div 24$ $= 0.208$
Step 2: Determine the mass of nitrogen	$m = n \times gfm$ $= 0.208 \times 28$ $= 5.82$ g

Reacting volumes

The volumes of reactant and product gases can be calculated from the number of moles of each reactant and product.

Remember: each gas will occupy the same volume under the same conditions of temperature and pressure.

Therefore, there is a direct relationship between number of moles and volume.

Examples

1. **What volume of carbon dioxide is formed when 20 cm³ of ethane is burnt in a plentiful supply of oxygen?**

$C_2H_6(g) + 3\frac{1}{2}O_2(g) \longrightarrow 2CO_2(g) + 3H_2O(l)$

Step 1: Determine the mole ratio	C_2H_6 : CO_2
Step 2: State volume ratio	1 mole : 2 mole 1 vol : 2 vol
Step 3: Calculate the volume	$20\ cm^3 \longrightarrow 2 \times 20$ $= 40\ cm^3$

2. **What volume of hydrogen is produced when 6.55 g of zinc reacts with excess hydrochloric acid?**

The molar volume of hydrogen is 21 litres mol^{-1}:

$Zn(s) + 2HCl(aq) \longrightarrow ZnCl_2(aq) + H_2(g)$

Step 1: Determine the mole ratio	$Zn:H_2$ 1 mole:1 mole
Step 2: Convert moles to mass and volume	65.5 g:21 litres
Step 3: Calculate the volume	$6.55 \longrightarrow \frac{6.55}{65.5} \times 21$ = 2.1 litres

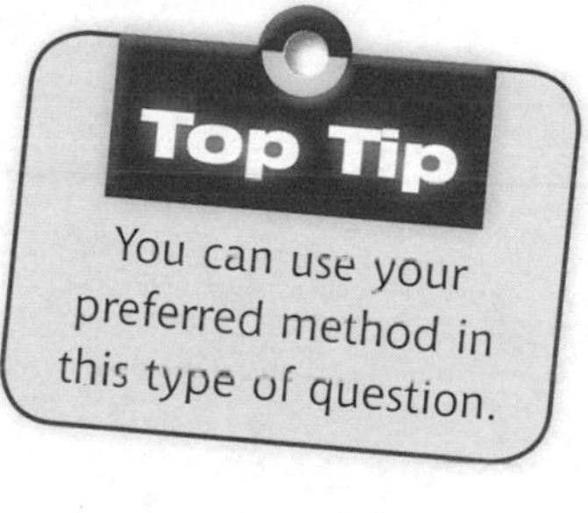

Questions

1. What volume of carbon dioxide is made when 200 g of calcium carbonate reacts with excess hydrochloric acid? (Molar volume of CO_2 = 22 litres mol^{-1})

   $CaCO_3(s) + 2HCl(aq) \longrightarrow CaCl_2(aq) + CO_2(g) + H_2O(l)$

2. What is the total volume of gases produced when 10 cm³ of propane burns in 80 cm³ oxygen?

   $C_3H_8(g) + 5O_2(g) \longrightarrow 3CO_2(g) + 4H_2O(l)$

Petrol

What is petrol?

Petrol is probably the best known of all products from crude oil. Crude oil is **fractionally distilled** to separate the fractions, and petrol is made by **reforming** the naphtha fraction. Petrol is a blend of hydrocarbons that contain between 5 and 10 carbons in their chains. The blend depends on the prevailing temperatures. In summer, petrol will contain fewer volatile hydrocarbons. This stops the petrol from evaporating too quickly. In winter, it will contain more volatile hydrocarbons, which will evaporate more quickly in cold weather.

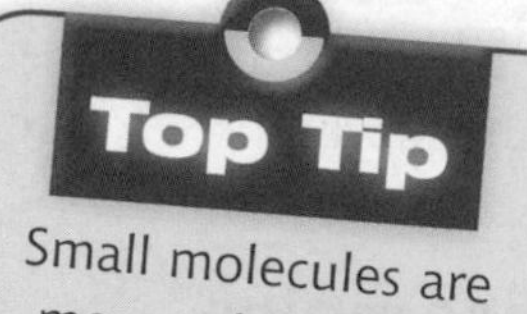

Small molecules are more volatile than larger molecules.

Petrol engine

Inside the engine, the combustion of petrol will produce hot gases that expand against parts of the engine and cause them to move.

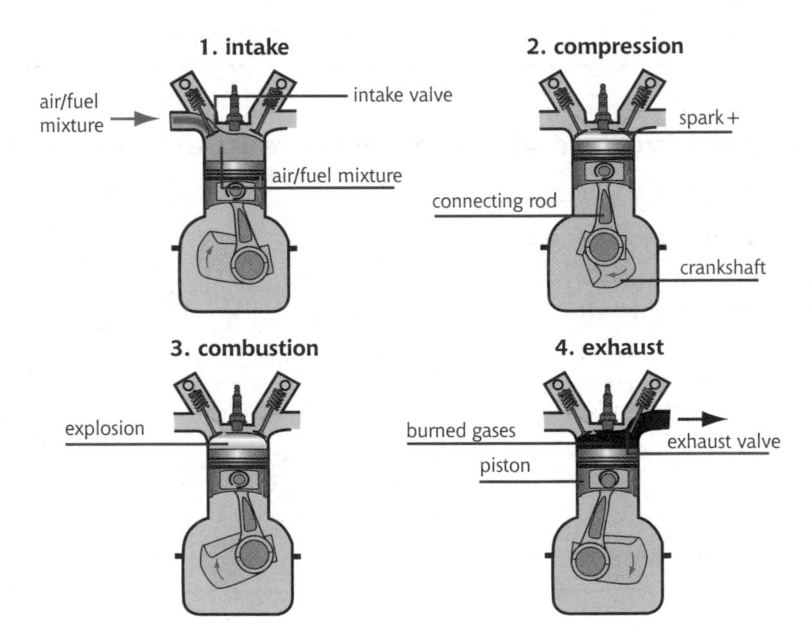

Chemically, when petrol is burnt it produces waste gases. Carbon dioxide, carbon monoxide, nitrogen oxides and water vapour are pushed out through the exhaust system.

Inside the engine, combustion of the petrol–air mixture is triggered by a spark. However, under certain conditions the fuel can auto-ignite spontaneously and out of sequence. This is called 'knocking' or 'auto ignition'. The noise associated with it is called 'knock' or 'tinkling'. Knocking is undesirable not just because of the noise but because it reduces engine performance. The octane number of a fuel is a measure of its ability to resist knocking. A fuel with a high octane number is less likely to cause knocking and a fuel with a low octane number will be more likely to cause knocking.

If petrol were used as it is produced straight from the fractionating column, it would have a very low octane number and would be unsuitable for its purpose.

To prevent knocking, additives may be added. In the past, small amounts of tetraethyllead, $Pb(C_2H_5)_4$ were used. However this was found to be extremely damaging to the environment and so unleaded petrol is now used.

There are 3 main ways to increase the burning efficiency of unleaded petrol:

1. Use alkanes with a high degree of branching instead of straight-chain alkanes.
2. Use aromatic hydrocarbons
3. Use cycloalkanes.

Reforming

Reforming alters the arrangement of atoms in molecules without necessarily changing the number of carbon atoms per molecule. As a result of the reforming process, petrol contains branched-chain alkanes, cycloalkanes and aromatic hydrocarbons as well as straight-chain alkanes.

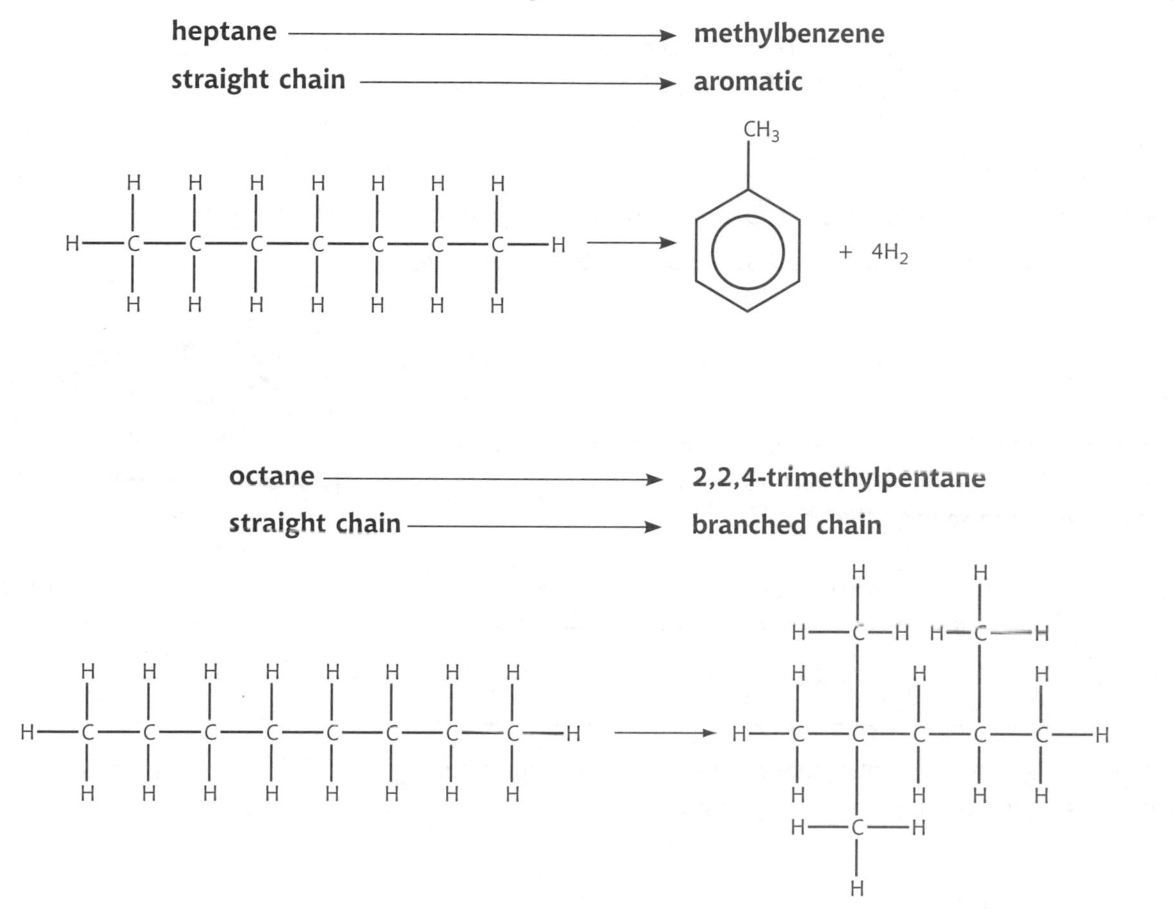

These molecules have much higher octane numbers and will therefore burn more smoothly and efficiently in the engine.

Questions

1. Name the process which alters the arrangement of carbon atoms in hydrocarbon molecules.
2. Which fraction is petrol obtained from?
3. What size of molecules should be added to winter-blend petrol to improve engine performance?

Alternative fuels

Alternative fuels are simply vehicle fuels that do not depend on petrol or diesel. The 2 main advantages are:

1. They greatly reduce harmful exhaust emissions.
2. They do not use up a finite resource – naphtha – which is obtained from crude oil.

Biofuels

Biofuels are fuels which come from plant or animal matter. There are 2 main biofuels: ethanol and biogas.

1. Ethanol

As a fuel, ethanol is produced extensively in Brazil. The carbohydrate sucrose, found in sugar cane, can be fermented to produce ethanol. Car engines can be adapted to run on a mixture of petrol and ethanol or purely ethanol.

The **advantages** of ethanol are:

- Sugar cane is an easy and profitable plant to grow and can be classed as renewable.
- Ethanol burns more efficiently and produces less carbon monoxide than petrol.

2. Biogas

Biogas is mainly made up of 50–80% methane. The rest of the gas is mainly made up of carbon dioxide.

It is made from the fermentation of plant and animal matter under anaerobic conditions. Bacteria act on waste materials such as livestock manure, livestock entrails, landfill-site waste and sewage sludge to form biogas.

'Anaerobic condition' means without oxygen.

Methanol

Methanol (CH_3OH) is a clear liquid alcohol that can be produced from natural gas, coal, crude oil and biomass crops such as wood. Its main synthesis route is by passing synthesis gas over a heated catalyst.

$$CO(g) + 2H_2(g) \longrightarrow CH_3OH(l)$$

synthesis gas — methanol

The **advantages** associated with using methanol compared to petrol are:

- It burns completely – producing fewer soot particles and less carbon monoxide.
- It is less volatile than petrol – that is, less likely to explode.

The **disadvantages** associated with using methanol compared to petrol are:

- 1 gallon of methanol produces half the energy of 1 gallon of petrol – giving fewer miles to the gallon.
- It is extremely toxic.
- Methanol absorbs water, which will clog filters and corrode engine parts, reducing engine performance.

Hydrogen

There are two common raw materials for hydrogen production: water and hydrocarbons such as methane.

- Hydrogen is produced by electrolysis of water. The major advantage of this is that there is an almost limitless supply of water. Using solar energy to produce electricity makes this environmentally friendly.
- Hydrogen can be produced when hydrocarbons react with steam. While this is a very simple process, it relies upon the Earth's finite reserves of hydrocarbons, such as methane, to make hydrogen.

The complete combustion of hydrogen is very clean.

$2H_2(g) + O_2(g) \longrightarrow 2H_2O(l)$

The main advantage of using hydrogen as a fuel is that this would reduce the build-up of carbon dioxide in the atmosphere, as only water is produced.

Currently hydrogen is used as a fuel in space rockets. Also, some vehicle manufacturers are developing hydrogen-powered engines.

The main difficulties in this are:

- hydrogen storage – in compressed or liquid form it needs a heavy, and therefore expensive, tank
- liquefying hydrogen is expensive
- hydrogen is highly flammable

Questions

1. State one advantage and two disadvantages of using methanol as an alternative fuel.
2. Explain the condition 'anaerobic'.
3. Why can ethanol be classed as a renewable fuel?

Hydrocarbons

Remember: a hydrocarbon is a compound which contains only the elements carbon and hydrogen. You need to be able to name and draw the full and shortened structural formula for:

- alkanes
- alkenes
- alkynes.

Alkanes

Alkanes have the general formula C_nH_{2n+2}

They have carbon-to-carbon single bonds, so the name must end in **-ane.**

The straight-chain molecule has all the carbon atoms in a continuous line, but the chain can be bent. For example, these two drawings show the same molecule, pentane.

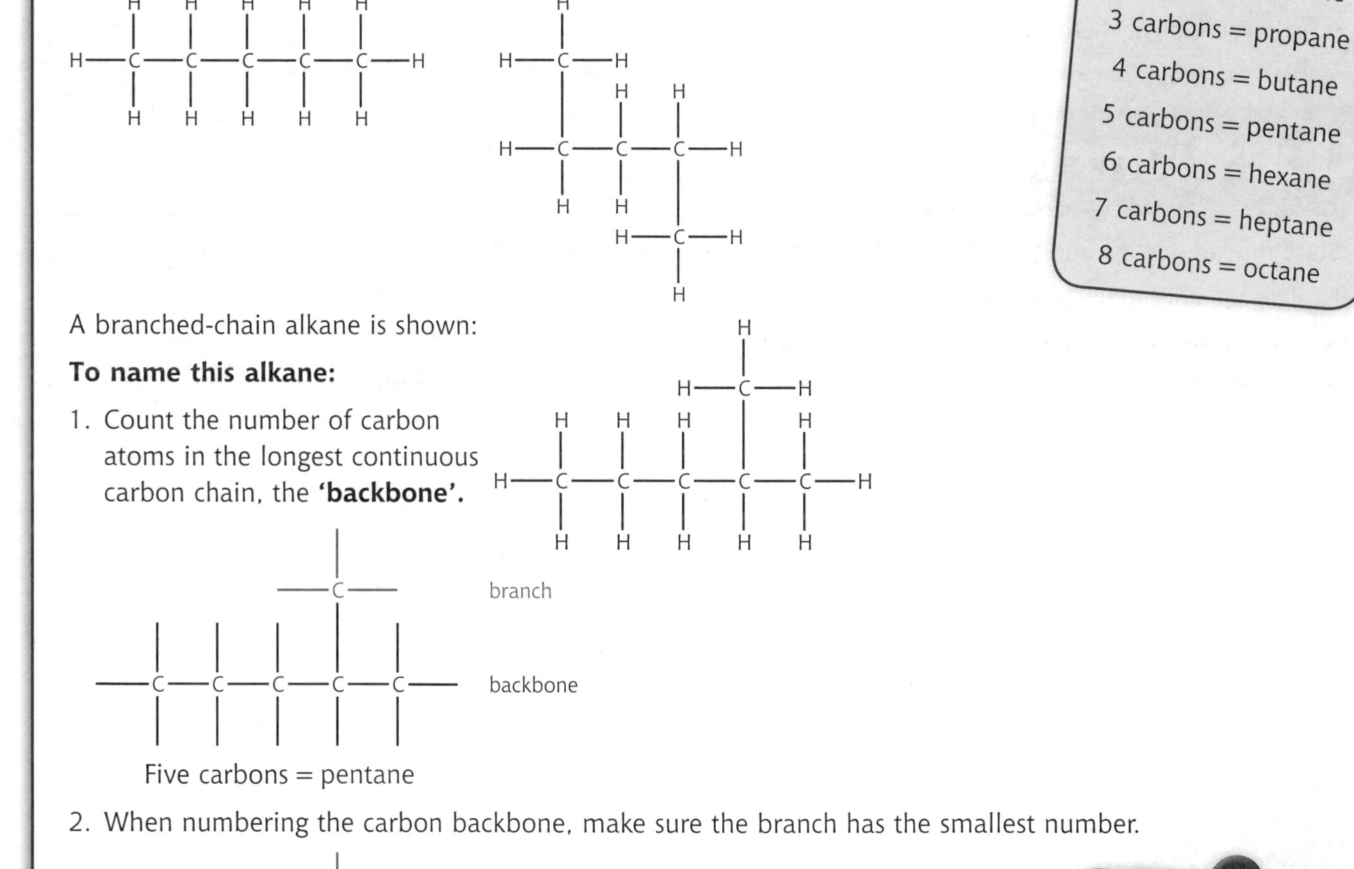

A branched-chain alkane is shown:

To name this alkane:

1. Count the number of carbon atoms in the longest continuous carbon chain, the **'backbone'.**

Five carbons = pentane

Top Tip

1 carbon = methane
2 carbons = ethane
3 carbons = propane
4 carbons = butane
5 carbons = pentane
6 carbons = hexane
7 carbons = heptane
8 carbons = octane

2. When numbering the carbon backbone, make sure the branch has the smallest number.

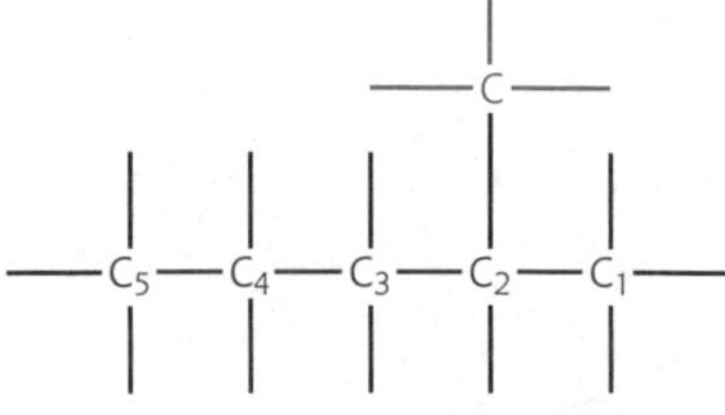

3. The branch is on position 2 and it contains 1 carbon atom, CH_3.

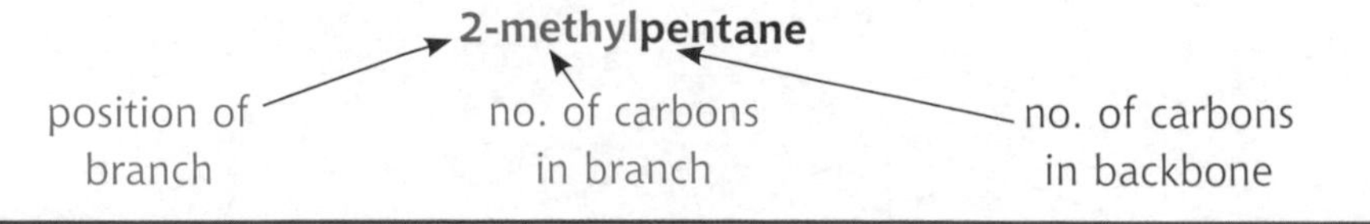

Top Tip

1 carbon, CH_3 = methyl
2 carbons, C_2H_5 = ethyl
3 carbons, C_3H_7 = propyl

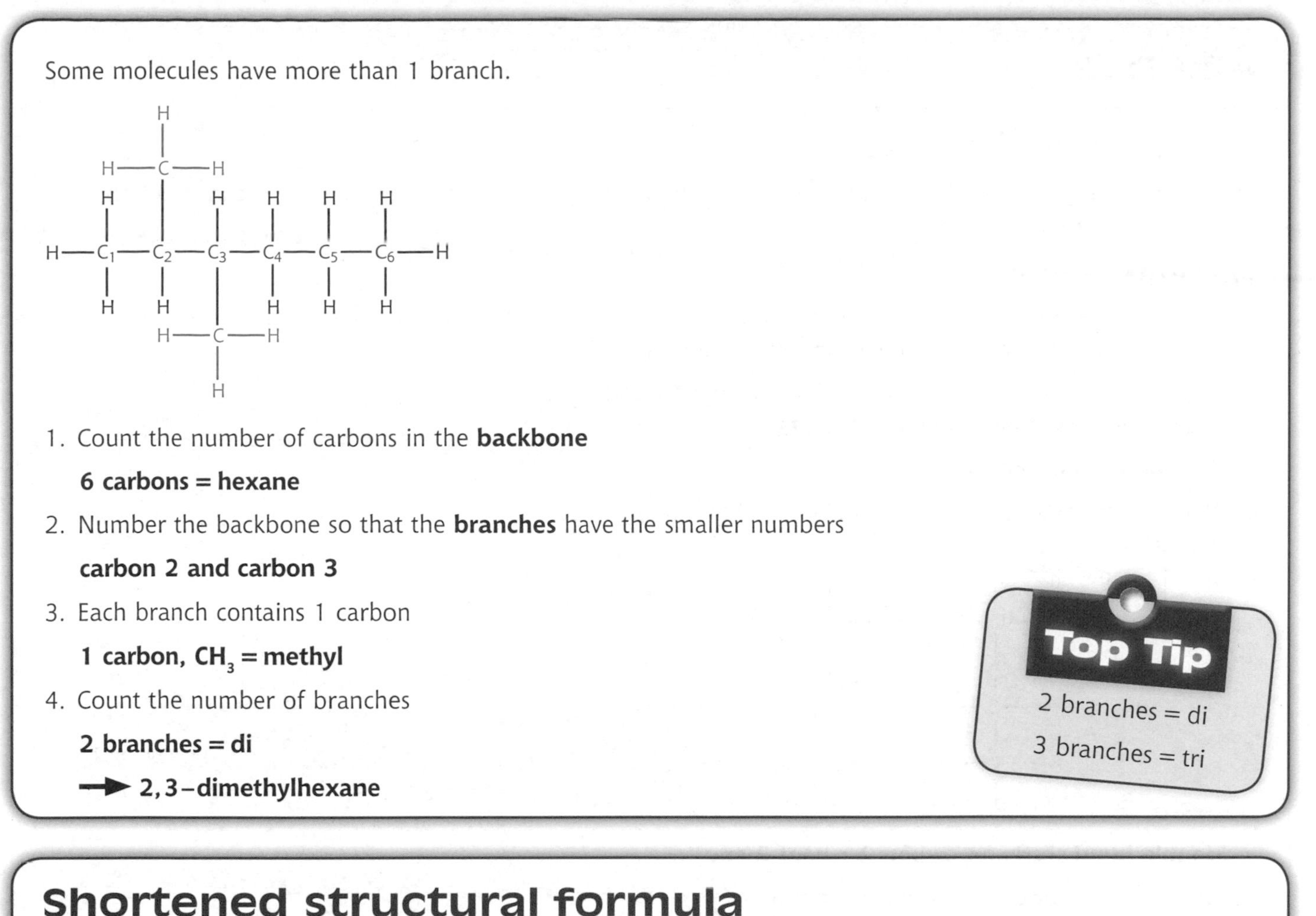

Some molecules have more than 1 branch.

1. Count the number of carbons in the **backbone**

   6 carbons = hexane
2. Number the backbone so that the **branches** have the smaller numbers

   carbon 2 and carbon 3
3. Each branch contains 1 carbon

   1 carbon, CH_3 = methyl
4. Count the number of branches

   2 branches = di

   → 2,3–dimethylhexane

Top Tip

2 branches = di

3 branches = tri

Shortened structural formula

This type of formula doesn't show any bonds.

Example 1: 2,2,3-trimethylbutane

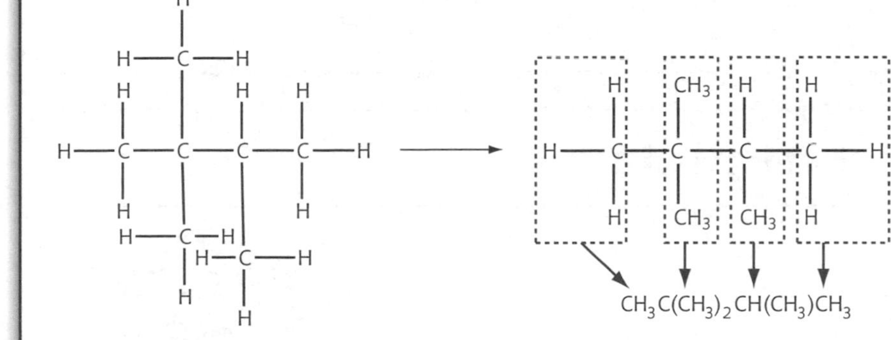

Example 2: 3-ethyl,5-methyloctane

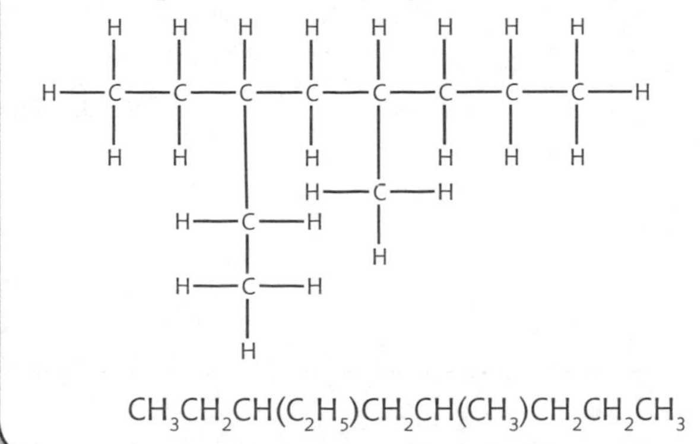

Alkenes

The general formula for alkenes is C_nH_{2n}

Their functional group is the carbon to carbon double bond, C═C, and this is the site of reactions. All alkene names end in **-ene**.

Naming

Follow the same rules as for alkanes except now the functional group, in this case the **carbon to carbon double bond, C═C, must have the smallest number**.

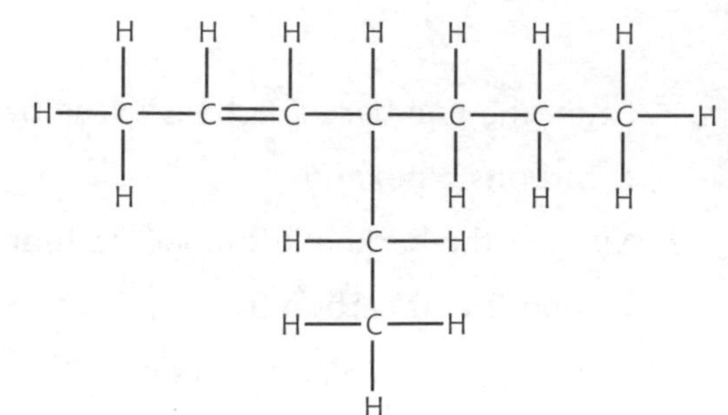

1. Count the number of carbons in the longest continuous carbon chain, the **backbone**, containing the C═C double bond.

   7 carbons = heptene

2. Number the carbons so that the double bond has the smallest number.

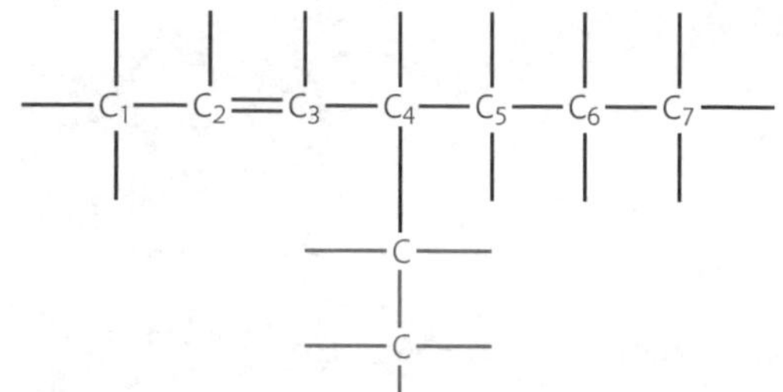

   double bond starts on carbon 2 = hept-2-ene

3. Number the **branch** – branch is on carbon 4.

   Identify the type of branch – 2 carbons, C_2H_5 = ethyl

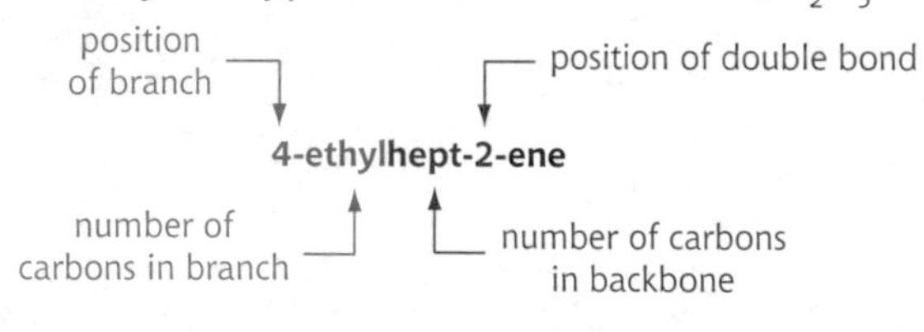

Shortened structural formula

$CH_3CH(CH_3)CH_2CH(CH_3)CHCH_2$

To name this structure, draw the full structural formula.

Be careful to count the number of bonds around each carbon atom to see where the double bond goes – remember carbon has 4 bonds.

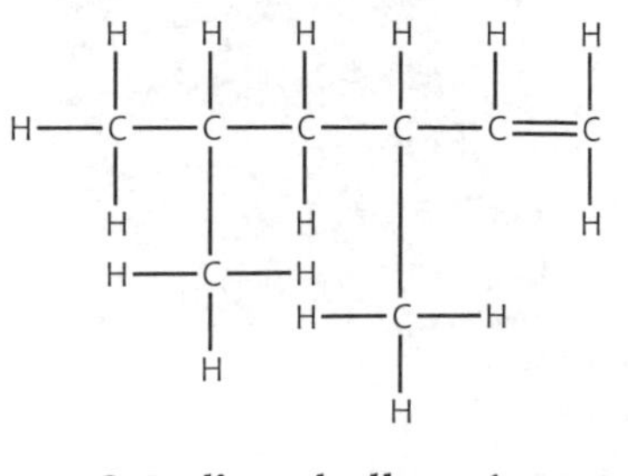

3,5-dimethylhex-1-ene

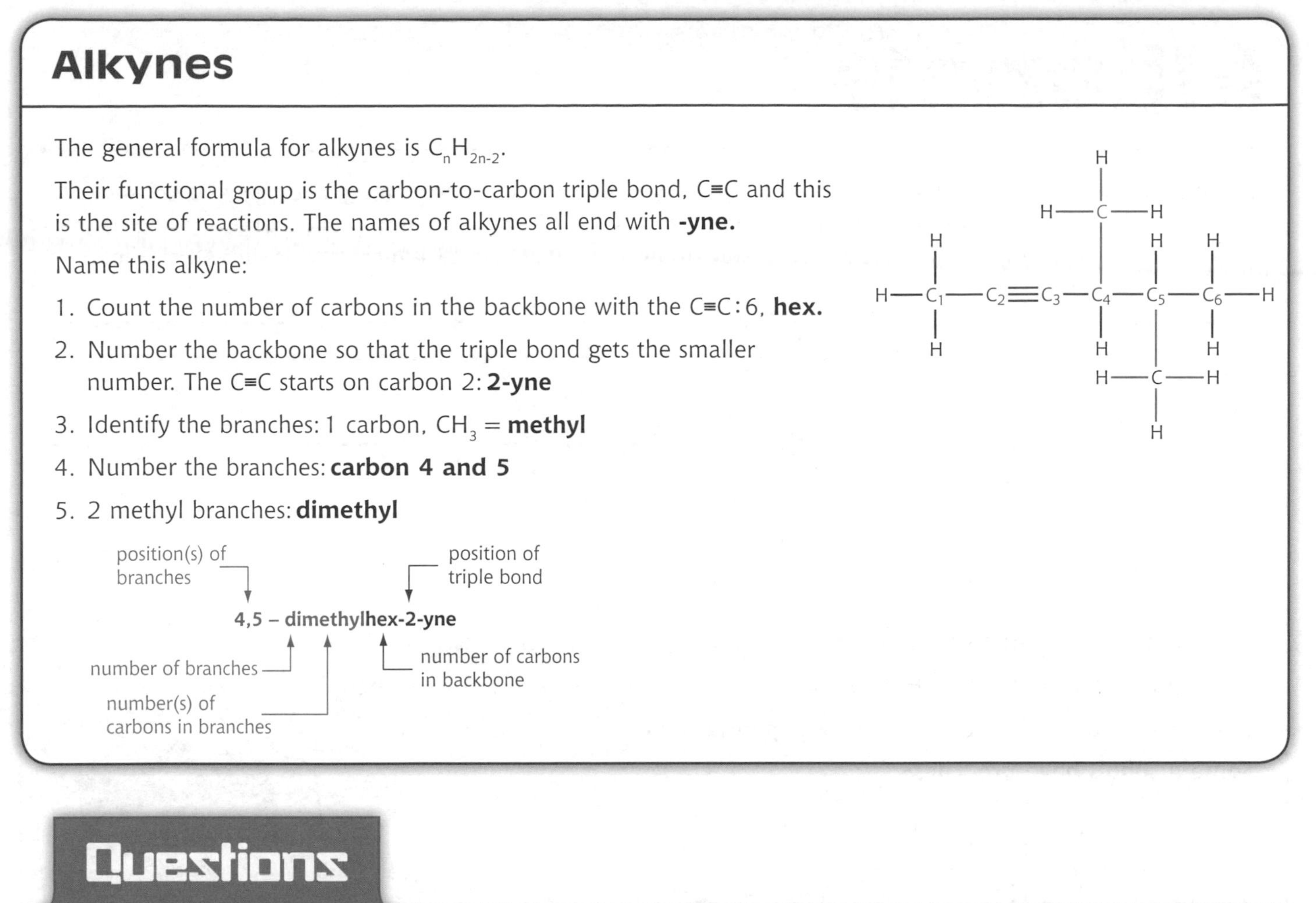

Alkynes

The general formula for alkynes is C_nH_{2n-2}.

Their functional group is the carbon-to-carbon triple bond, C≡C and this is the site of reactions. The names of alkynes all end with **-yne.**

Name this alkyne:

1. Count the number of carbons in the backbone with the C≡C: 6, **hex.**
2. Number the backbone so that the triple bond gets the smaller number. The C≡C starts on carbon 2: **2-yne**
3. Identify the branches: 1 carbon, CH_3 = **methyl**
4. Number the branches: **carbon 4 and 5**
5. 2 methyl branches: **dimethyl**

4,5 – dimethylhex-2-yne

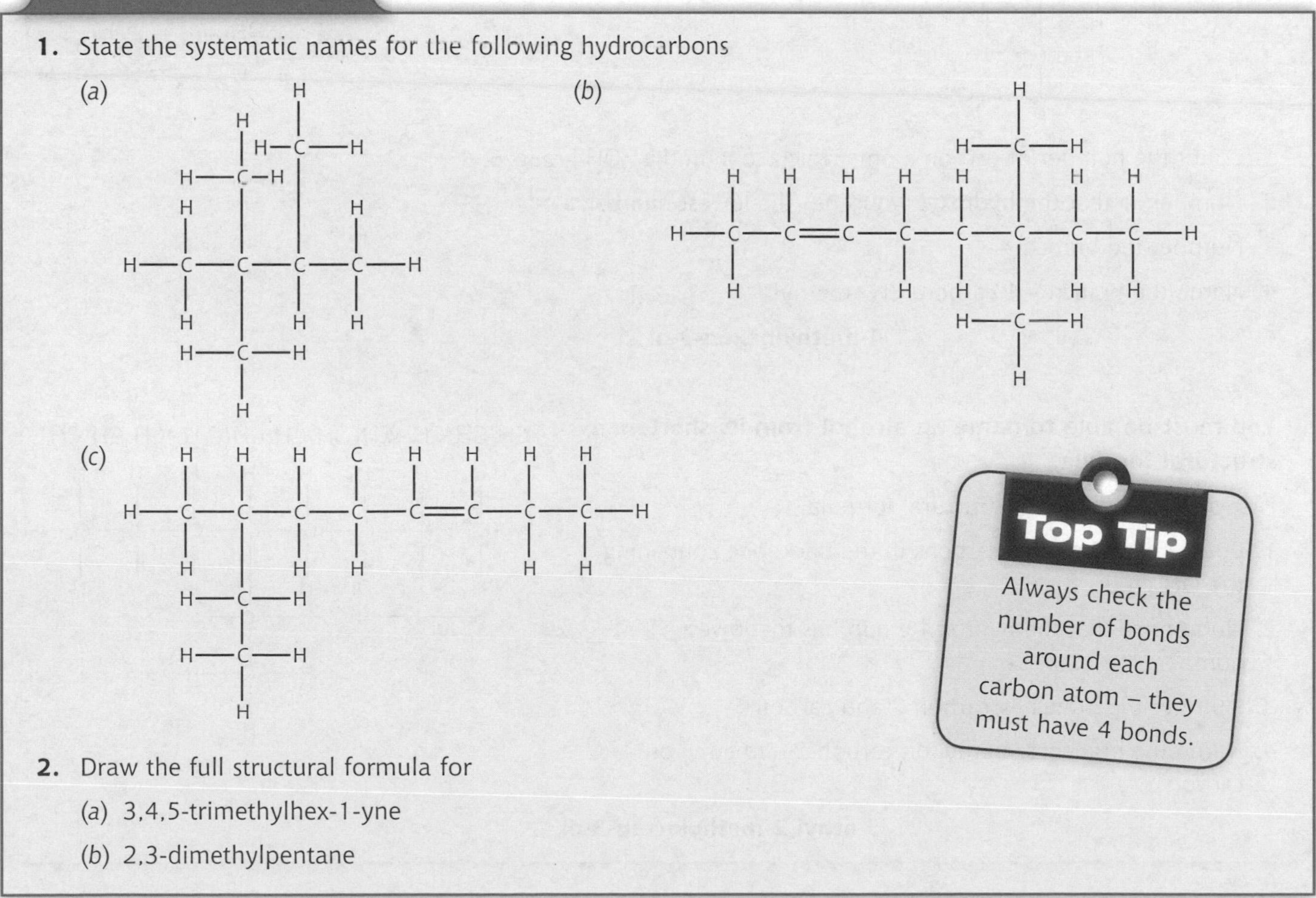

Questions

1. State the systematic names for the following hydrocarbons

   (a)

   (b)

   (c)

2. Draw the full structural formula for

   (a) 3,4,5-trimethylhex-1-yne

   (b) 2,3-dimethylpentane

Top Tip

Always check the number of bonds around each carbon atom – they must have 4 bonds.

Alkanols

Some families of organic compounds are based on alkane structures but one or more hydrogens have been replaced by a different group. Alkanols are a homologous series of alcohols based on the corresponding parent alkanes. An alcohol contains a hydroxyl functional group, –OH. Reactions take place at this site. All alcohol names end in **-ol.**

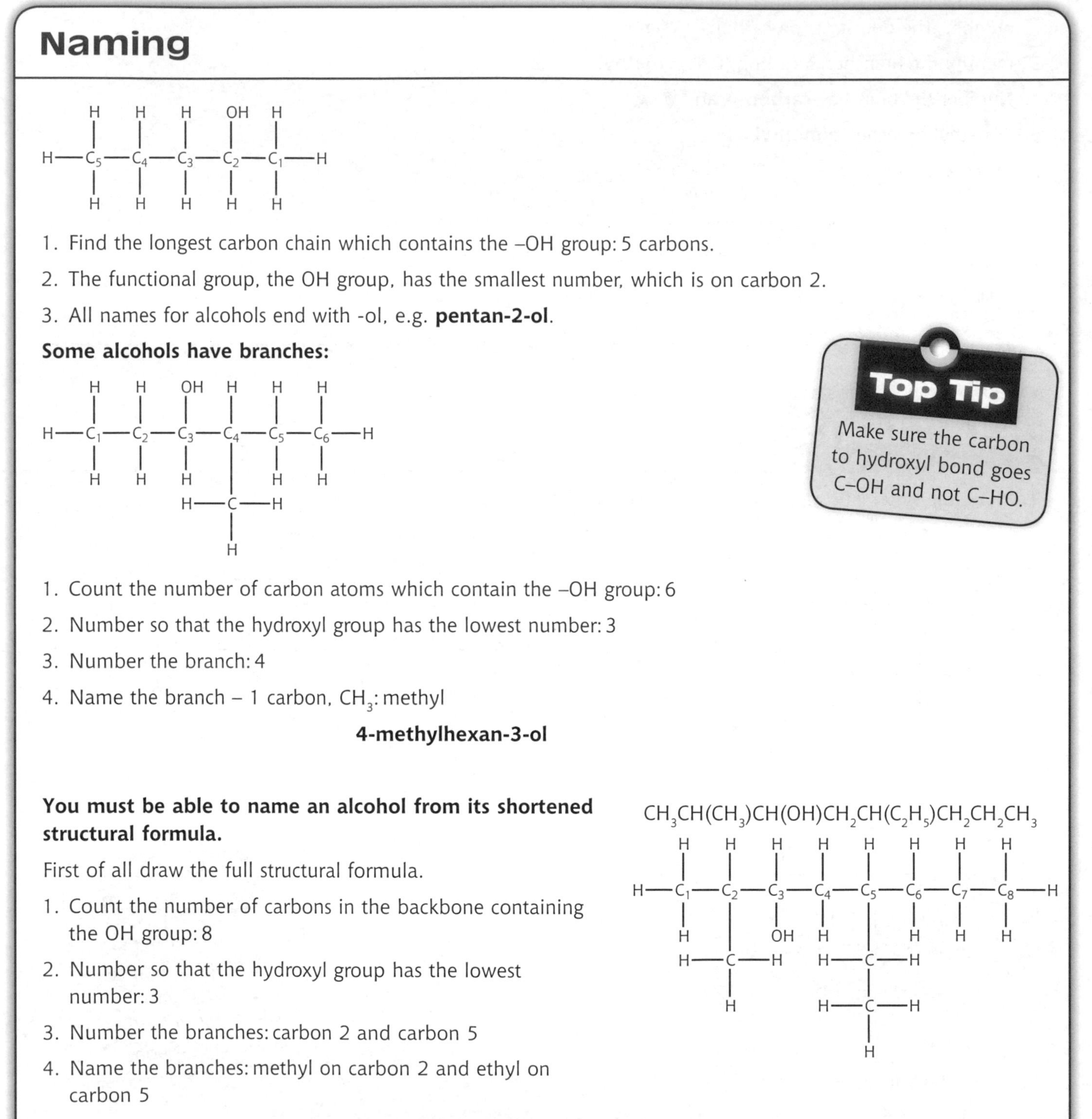

Naming

1. Find the longest carbon chain which contains the –OH group: 5 carbons.
2. The functional group, the OH group, has the smallest number, which is on carbon 2.
3. All names for alcohols end with -ol, e.g. **pentan-2-ol**.

Some alcohols have branches:

1. Count the number of carbon atoms which contain the –OH group: 6
2. Number so that the hydroxyl group has the lowest number: 3
3. Number the branch: 4
4. Name the branch – 1 carbon, CH_3: methyl

4-methylhexan-3-ol

Top Tip

Make sure the carbon to hydroxyl bond goes C–OH and not C–HO.

You must be able to name an alcohol from its shortened structural formula.

$CH_3CH(CH_3)CH(OH)CH_2CH(C_2H_5)CH_2CH_2CH_3$

First of all draw the full structural formula.

1. Count the number of carbons in the backbone containing the OH group: 8
2. Number so that the hydroxyl group has the lowest number: 3
3. Number the branches: carbon 2 and carbon 5
4. Name the branches: methyl on carbon 2 and ethyl on carbon 5

5-ethyl,2-methyloctan-3-ol

Classification

Alcohols can be classified as primary, secondary or tertiary. It is important to be able to classify alcohols, so you can predict how they will oxidise.

Primary alcohol

Find the functional group. Remember it is called a hydroxyl group, -OH

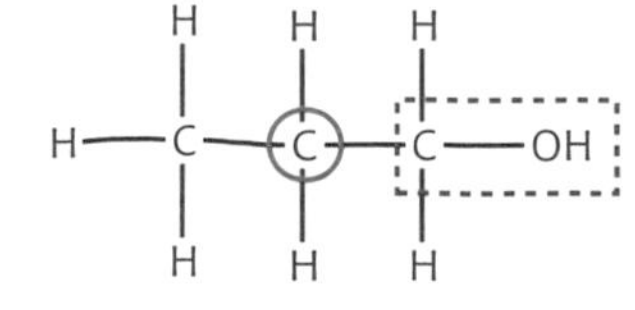

Count the number of carbons that are bonded to the carbon which is bonded to the hydroxyl group.

⇒ **1** carbon means that this is a **primary alcohol**

This alcohol is called propan-1-ol.

Secondary alcohol

Again, find the hydroxyl group and count the number of carbons that are bonded to the carbon which is directly attached to the hydroxyl group.

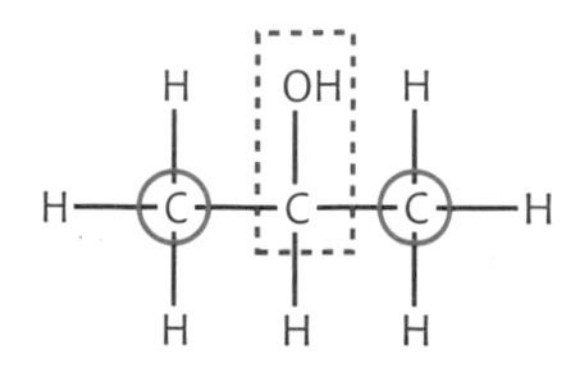

⇒ **2** carbons means that this is a **secondary alcohol**

This alcohol is called propan-2-ol.

Tertiary alcohol

Follow the rule and count the carbons.

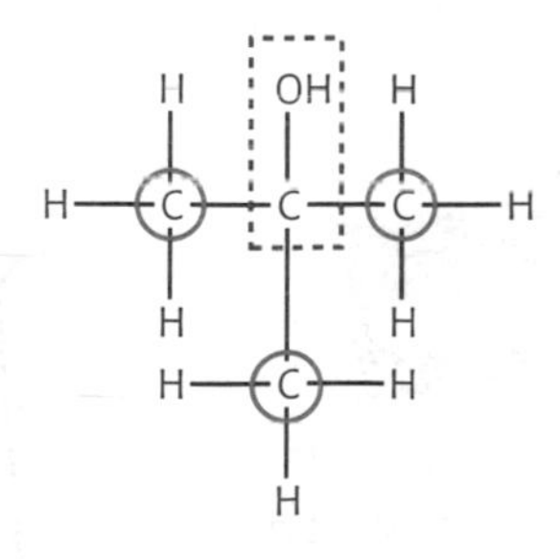

⇒ **3** carbons means this is a **tertiary alcohol**

This alcohol is called 2-methylpropan-2-ol.

Primary alcohols will always have the OH group at the end of the molecule.

Secondary alcohols will have the OH group in the middle of the molecule.

Tertiary alcohols will have the OH group in the middle of the molecule directly opposite a branch.

Questions

1. Draw the full structural formula for the following alcohols and state what class of alcohol they are:

(*a*) pentan-3-ol

(*b*) 3-methylbutan-2-ol

(*c*) 4,4,5-trimethylheptan-1-ol

Esters

Esters are a homologous series that can be identified by their carboxylate functional group, more commonly called the **ester link.**

Naming

Esters are formed by the **condensation** reaction between an alkanoic acid and an alcohol. All esters end in **-oate.**

ALCOHOL + ALKANOIC ACID ⇌ ESTER + WATER

ethanol + propanoic acid ⇌ ethyl propanoate + water

The first part of the ester's name comes from the alcohol:

alcohol	1st part of ester name
methanol	methyl
ethanol	ethyl
propanol	propyl
butanol	butyl

A condensation reaction joins molecules together by the elimination of a small molecule, most commonly water.

The second part of the ester's name comes from the alkanoic acid:

alkanoic acid	2nd part of ester name
methanoic	methanoate
ethanoic	ethanoate
propanoic	propanoate
butanoic	butanoate

Making esters

Ethyl propanoate is formed by the reaction of the **hydroxyl** group from the alcohol and the **carboxyl** group from the alkanoic acid.

Ethanol + propanoic acid ⇌ ethyl propanoate + water

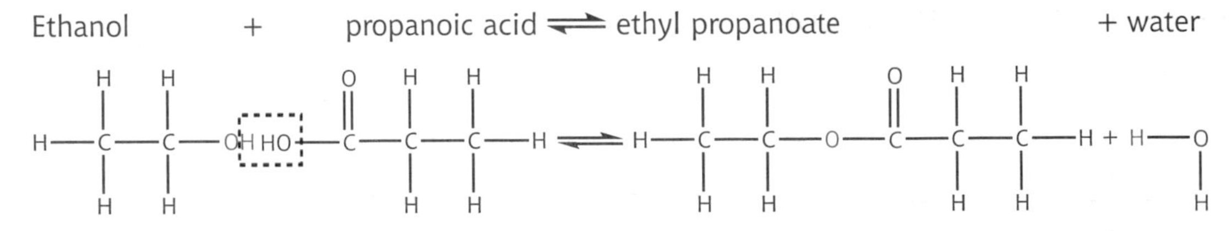

This reaction is commonly called esterfication. Water is always made in this reaction.

Making esters is a reversible reaction – you will not obtain 100% of the theoretical yield.

Naming an ester from a full structural formula

1. Find the ester link.
2. Draw a line that splits the C–O bond in the ester link.
3. The 'half' of the molecule which contains the O–C bond belongs to the parent alcohol – count the number of carbon atoms.

   3 carbons = propanol

Esters are used as flavourings, perfumes and solvents.

4. The 'half' of the molecule that contains the C=O bond belongs to the parent acid – count the number of carbon atoms.

   4 carbons = butanoic acid

   The ester is **propyl butanoate.**

3. parent alcohol

4. parent alkanoic acid

Naming an ester from a shortened structural formula

The example here is CH_3CH_2OOCH

1. Draw out its full structural formula:
2. Find the ester link.
3. Draw a line which splits the link.
4. The 'half' which contains the O–C bond belongs to the alcohol – count the number of carbon atoms.

   2 carbons = ethanol

5. The 'half' which contains the C=O bond belongs to the acid – count the number of carbon atoms.

   1 carbon = methanoic acid

   The ester is **ethyl methanoate.**

Breaking esters

Esters can be broken down in a hydrolysis reaction. This is a reversible reaction.

propyl ethanoate + water ⇌ propanol + ethanoic acid

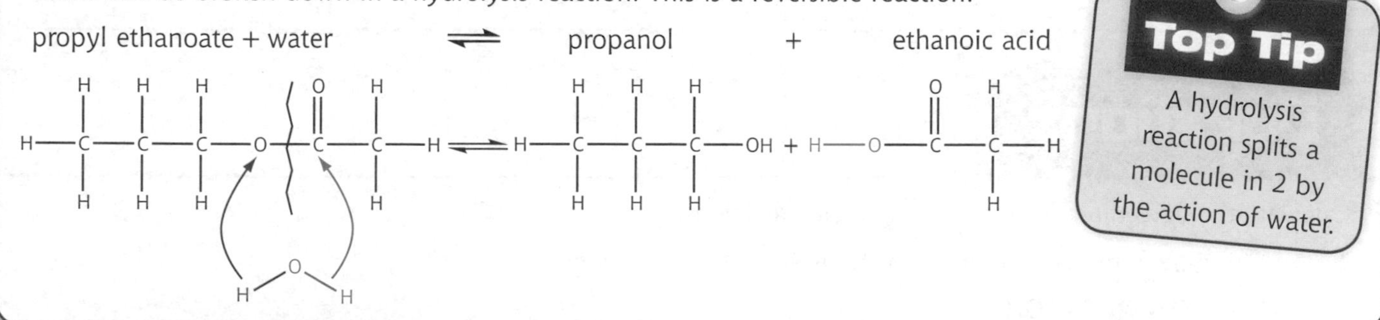

Top Tip

A hydrolysis reaction splits a molecule in 2 by the action of water.

Questions

1. Name and draw the full structural formula of the ester made from the following reactants
   (*a*) hexan-1-ol + butanoic acid
   (*b*) HCOOH + $CH_3CH_2CH_2OH$
2. Name and draw the full structural formulae of the alkanol and alkanoic acid that would be formed from the breakdown of the following esters
   (*a*) ethyl ethanoate
   (*b*) methyl octanoate

Percentage yield calculations

It is important for chemists to be able to calculate the % yield of a reaction.

This is a 2-step process.

Step 1: Calculate the theoretical yield

This is the maximum amount that could be produced in a reaction. Reactions rarely produce 100% products, mainly due to the fact that they are commonly reversible.

What is the theoretical yield of ethyl ethanoate that can be produced when 4·6 g of ethanol reacts with excess ethanoic acid?

$C_2H_5OH + CH_3COOH \rightleftharpoons CH_3COOC_2H_5 + H_2O$

Identify the mole ratio	C_2H_5 1 mole of ethanol	:	$CH_3COOC_2H_5$ 1 mole of ethyl ethanoate
Change moles to mass	46 g	:	88 g
Calculate mass of ester	4.6 g	⟶	$\frac{4.6 \times 88}{46} = 8.8$ g

The theoretical yield is 8·8 g.

Step 2: Calculate the % yield

After carrying out the reaction, a student obtained 6·2 g of ethyl ethanoate.

To calculate the % yield, use the following equation:

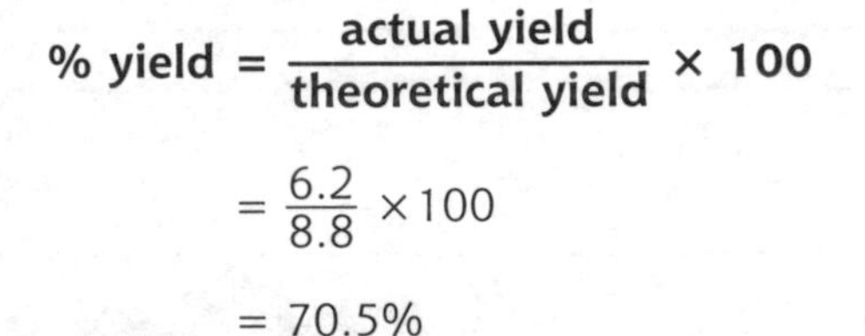

$$\textbf{\% yield} = \frac{\textbf{actual yield}}{\textbf{theoretical yield}} \times \textbf{100}$$

$$= \frac{6.2}{8.8} \times 100$$

$$= 70.5\%$$

Questions

Calculate the percentage yield of the following reactions:

1. 14 g of ethene reacts with excess water to form 13 g of ethanol

$C_2H_4 + H_2O \rightleftharpoons C_2H_5OH$

2. 60 g of hydrogen reacts with excess nitrogen to make 255 g of ammonia

$N_2 + 3H_2 \rightleftharpoons 2NH_3$

Aromatic hydrocarbons

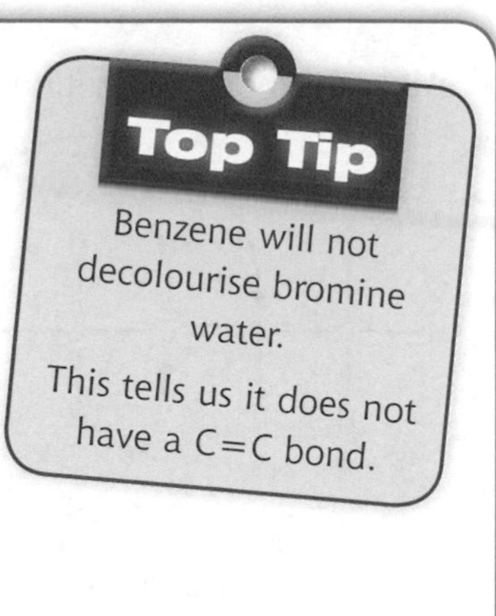

The simplest aromatic hydrocarbon is the colourless liquid, **benzene**.

It has the molecular formula C_6H_6 and it has a distinctive cyclic structural formula.

Its structural formula can be shown as:

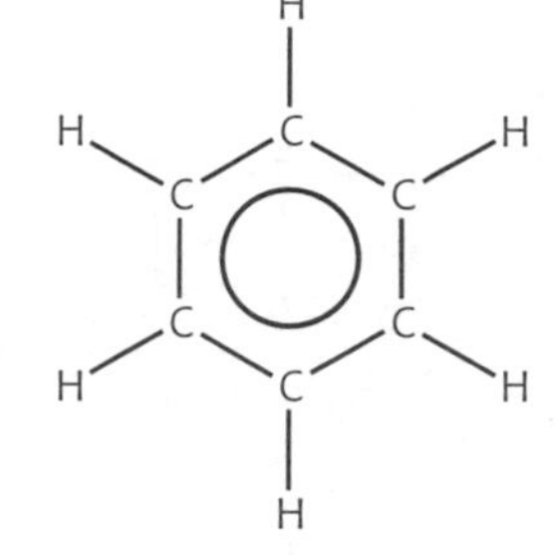

Structural analysis shows that it contains 6 carbon to carbon bonds of equal length and strength.

Each carbon atom has 4 outer electrons. 3 electrons are used in bonding – 2 to the adjacent carbons and 1 to the hydrogen.

This leaves 1 electron from each carbon atom to form an electron cloud above and below the cyclic ring. These electrons are free to move around the carbon ring and are said to be delocalised.

6 delocalised electrons

These delocalised electrons make the benzene ring very **stable**. Benzene will resist addition reactions. Commonly we see the benzene structure represented as:

A benzene ring in which one of the hydrogen atoms has been substituted by another group is known as the **phenyl group.**

The phenyl group has the formula $-C_6H_5$.

Examples of other aromatic compounds are shown in the table:

Name	Structure	Molecular formula
phenol	OH	C_6H_6O
phenylamine	NH_2	C_6H_7N
methylbenzene	CH_3	C_7H_8

Benzene and its related compounds are important feedstocks (starting materials), used to make detergents, explosives and germicides (e.g. TCP, dettol).

Addition reactions

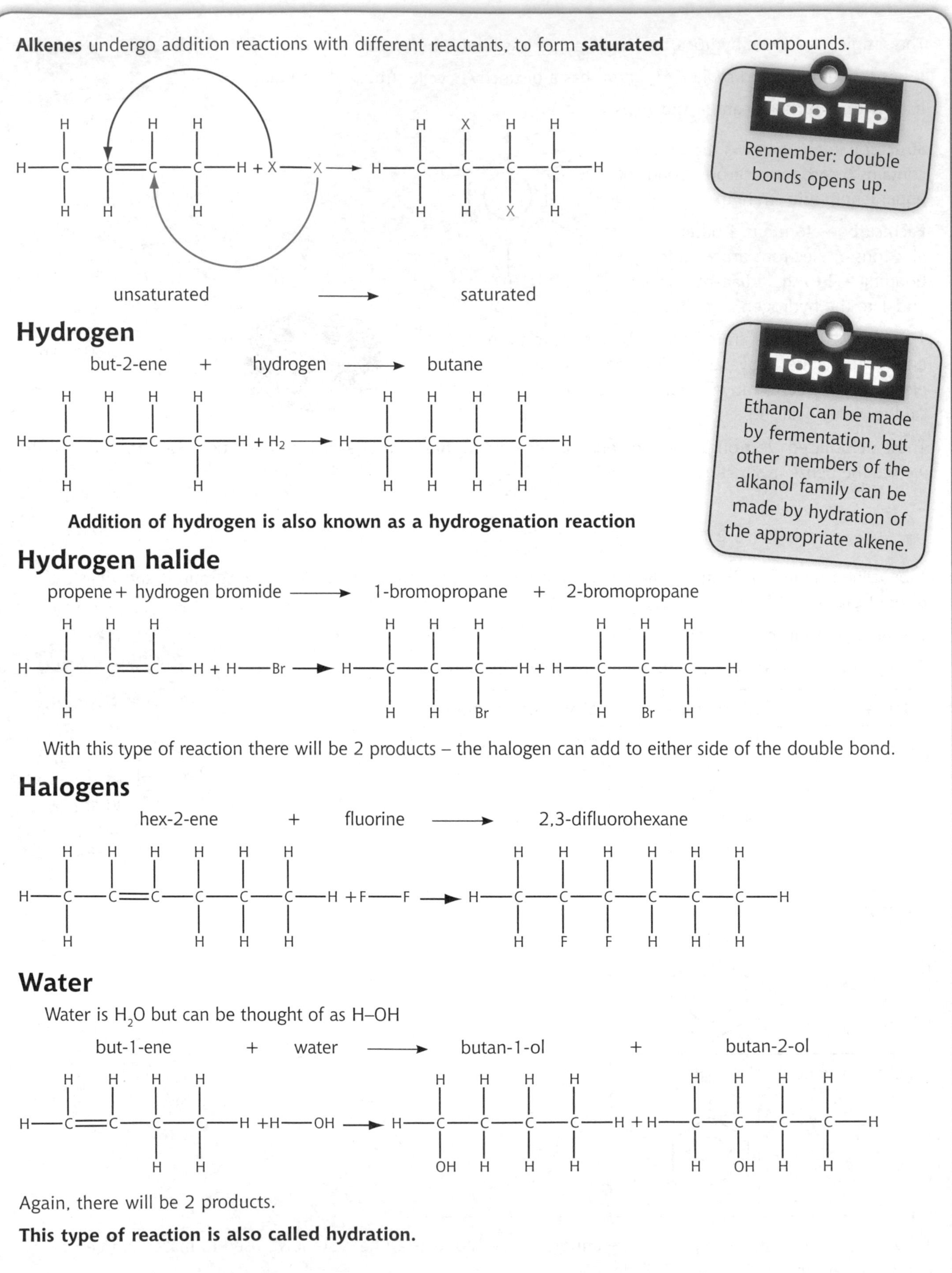

Alkenes undergo addition reactions with different reactants, to form **saturated** compounds.

unsaturated ⟶ saturated

> **Top Tip**
> Remember: double bonds opens up.

Hydrogen

but-2-ene + hydrogen ⟶ butane

$$H_3C{-}CH{=}CH{-}CH_3 + H_2 \rightarrow H_3C{-}CH_2{-}CH_2{-}CH_3$$

Addition of hydrogen is also known as a hydrogenation reaction

> **Top Tip**
> Ethanol can be made by fermentation, but other members of the alkanol family can be made by hydration of the appropriate alkene.

Hydrogen halide

propene + hydrogen bromide ⟶ 1-bromopropane + 2-bromopropane

With this type of reaction there will be 2 products – the halogen can add to either side of the double bond.

Halogens

hex-2-ene + fluorine ⟶ 2,3-difluorohexane

Water

Water is H_2O but can be thought of as H–OH

but-1-ene + water ⟶ butan-1-ol + butan-2-ol

Again, there will be 2 products.

This type of reaction is also called hydration.

Dehydration reaction

An alcohol can be converted to an alkene by dehydration. It is simply the opposite reaction to hydration.

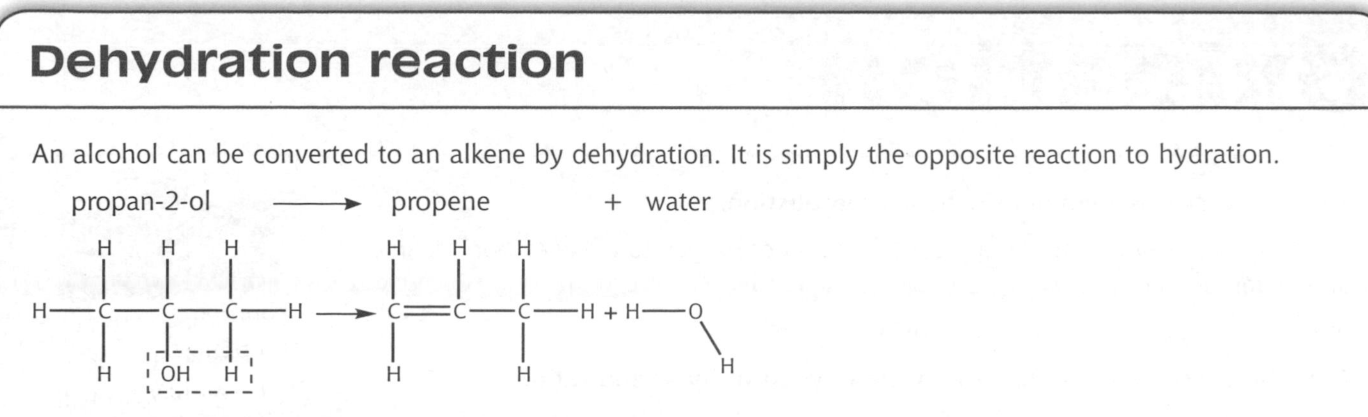

Alkynes will also undergo addition reactions. The reaction will take place in 2 stages to form a saturated product.

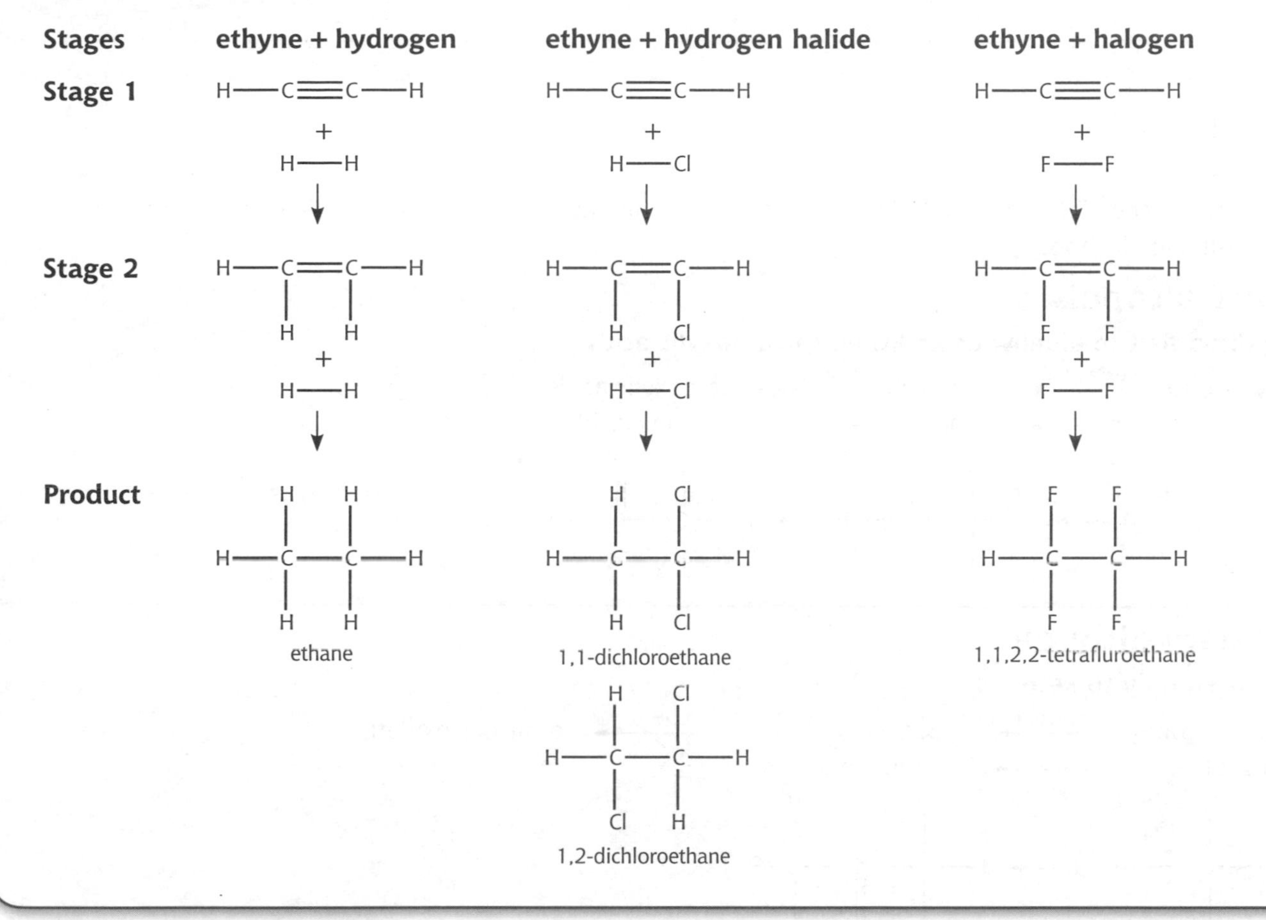

Questions

1. Name and draw the full structural formulae for the products of the following:
   (*a*) ethene + water
   (*b*) pent-2-ene + hydrogen iodide
   (*c*) but-1-ene + hydrogen
2. Name the chemical reaction which involves the addition of water.
3. Name the 2 products formed when ethyne reacts fully with hydrogen chloride.
4. How many products will be formed in the following addition reactions?
   (*a*) but-2-ene + water
   (*b*) pent-1-ene + hydrogen
   (*c*) hex-2-ene + hydrogen fluoride
   (*d*) hept-2-ene + water

Oxidation

The most complete form of oxidation is **combustion**.

For example, alcohols burn in a plentiful supply of oxygen to make carbon dioxide and water.

ethanol + oxygen ⟶ carbon dioxide + water

Oxidation results in an increase in the oxygen to hydrogen ratio.

Burning a carbon compound in a poor supply of oxygen will make the poisonous gas carbon monoxide.

Alcohols

Alcohols can be chemically oxidised. The product formed will depend on the **class** of alcohol. (See page 47 about classifying alcohols.)

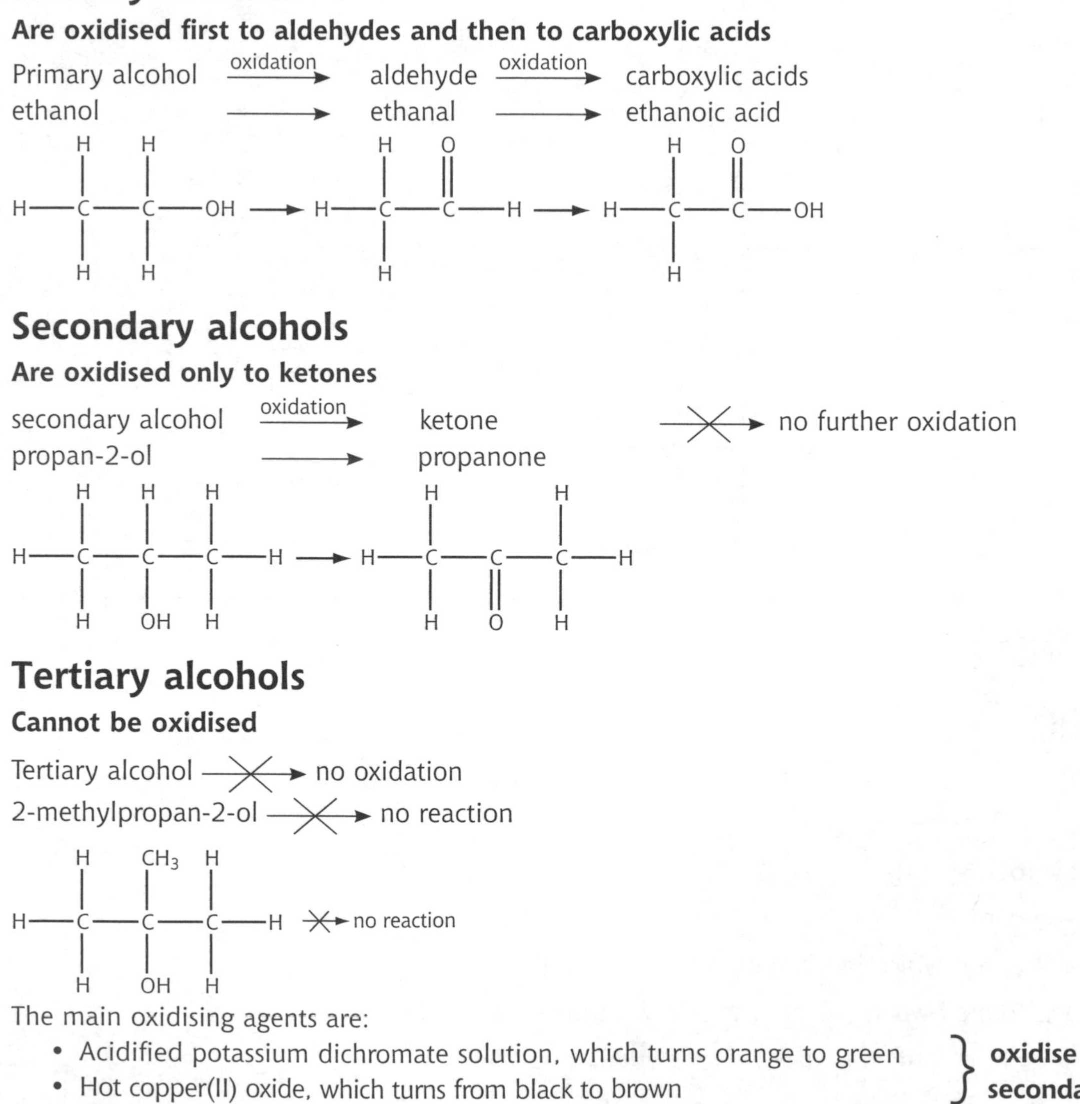

Primary alcohols

Are oxidised first to aldehydes and then to carboxylic acids

Primary alcohol —oxidation⟶ aldehyde —oxidation⟶ carboxylic acids

ethanol ⟶ ethanal ⟶ ethanoic acid

Secondary alcohols

Are oxidised only to ketones

secondary alcohol —oxidation⟶ ketone —✕⟶ no further oxidation

propan-2-ol ⟶ propanone

Tertiary alcohols

Cannot be oxidised

Tertiary alcohol —✕⟶ no oxidation

2-methylpropan-2-ol —✕⟶ no reaction

The main oxidising agents are:

- Acidified potassium dichromate solution, which turns orange to green
- Hot copper(II) oxide, which turns from black to brown

} **oxidise primary and secondary alcohols**

- Benedict's solution, which turns from blue to brick-red
- Tollen's reagent, which forms a silver mirror

} **oxidise aldehydes**

Reduction

Aldehydes and ketones can be reduced back to their corresponding alcohols using the reducing agent, lithium aluminium hydride.

Reduction results in a decrease in the oxygen to hydrogen ratio.

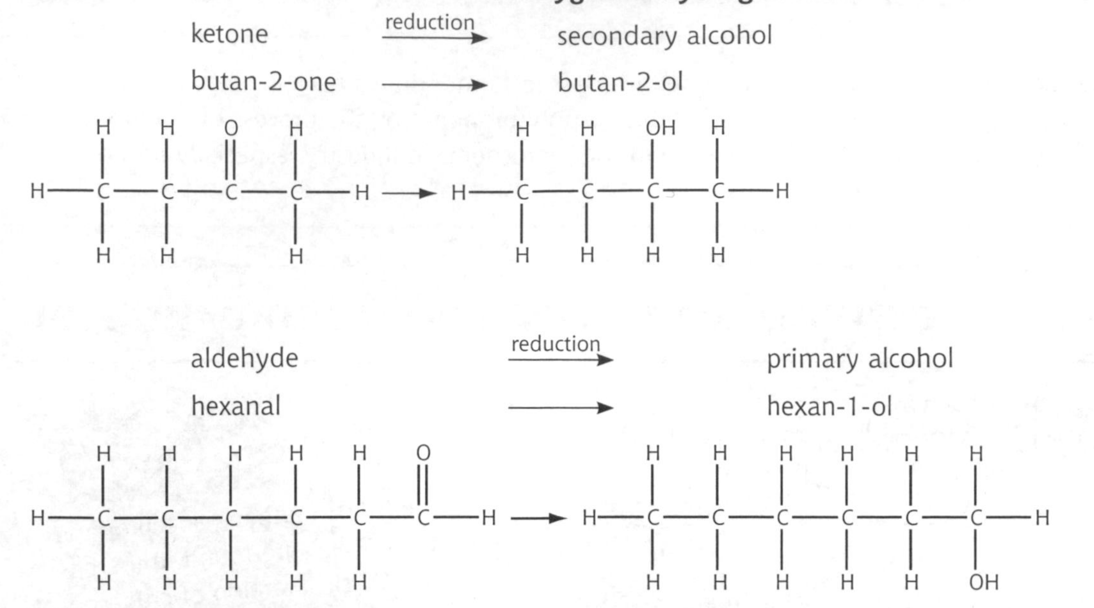

Questions

1. Name and draw the full structural formula of the products of the following **oxidation** reactions:
   (*a*) butan-2-ol
   (*b*) methanol
   (*c*) propan-2-ol
   (*d*) 3-methylpentan-3-ol
   (*e*) octanal
   (*f*) heptan-3-one

Remember: not all carbonyl compounds will oxidise.

2. Name and draw the shortened structural formula of the products of the following **reduction** reactions:
   (*a*) methanal
   (*b*) hexan-3-one
   (*c*) butan-2-one
   (*d*) hexanal
   (*e*) pentanoic acid

Early plastics and fibres 1

Addition polymers

Small unsaturated molecules, such as ethene and propene, are classed as monomers because they contain a C=C double bond. They can form the plastics poly(ethene) and poly(propene) by the process of addition polymerisation. Ethene is a feedstock of major importance in the petrochemical industry, especially in the manufacture of plastics.

Manufacture of ethene and propene from crude oil

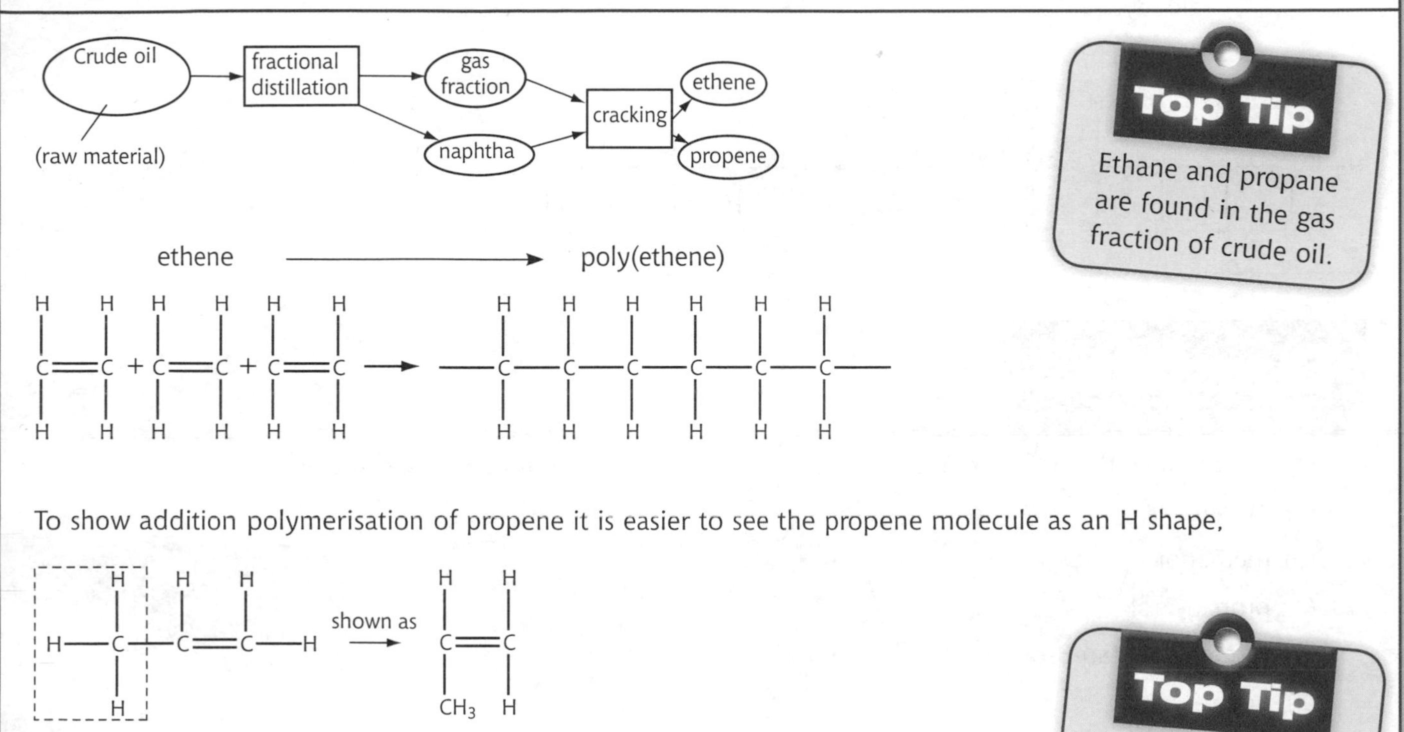

Top Tip

Ethane and propane are found in the gas fraction of crude oil.

To show addition polymerisation of propene it is easier to see the propene molecule as an H shape,

propene ⟶ poly(propene)

Top Tip

Remember to put the end bonds in the polymer chains.

Condensation polymers

Condensation polymers are made from monomers which have 2 functional groups per molecule.

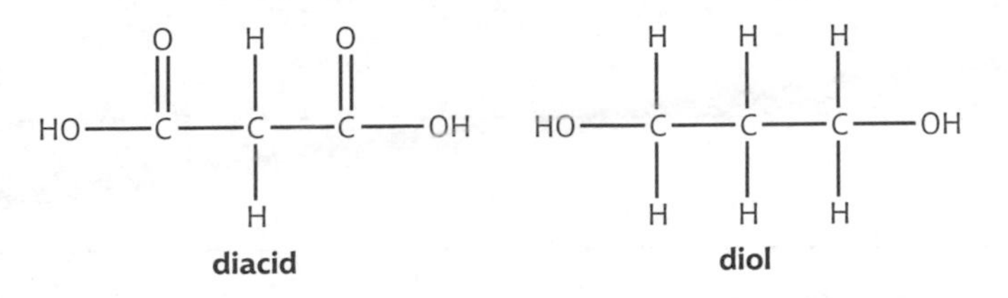

In a condensation reaction, molecules join together through the elimination of small molecules such as water. When the monomers have 2 functional groups, the polymer chain can grow in both directions.

Polyesters

These are examples of condensation polymers

Polyesters are a group of polymers which contain the ester functional group. The monomers must contain a diol and a diacid. A common polyester is terylene. It is also called poly(ethylene terephthalate), or PET for short.

PET can be made by the condensation reaction of benzene-1,4-dicarboxylic acid and ethane-1,2-diol.

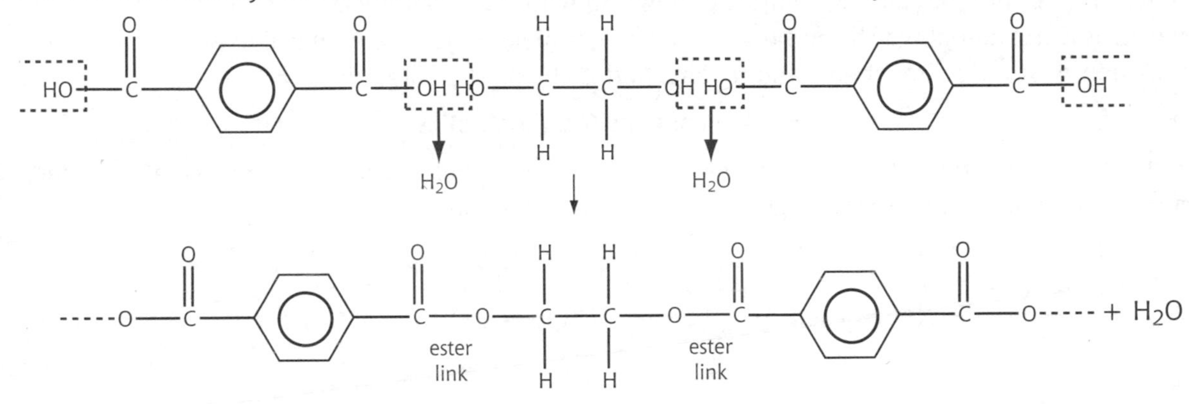

Resins and fibres

A polyester fibre is a thermoplastic linear polymer.

The polymer chains have weak van der Waals forces between the chains: these forces which can easily be broken.

A polyester resin has strong covalent cross-links between the polymer chains. This makes a rigid 3-D shape, which cannot be easily broken on heating. This makes it a **thermosetting** plastic.

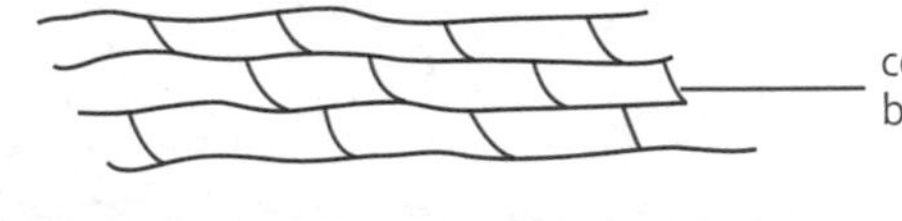

Questions

1. Why do resins have higher melting points than fibres?
2. Name the two types of monomer which make a polyester.

Early plastics and fibres 2

Polyamides

Polyamides are formed by the condensation reaction of a di-amine and a di-acid.

Amines contain the amino functional group $-NH_2$ -----N—H (with N above)

Carboxylic acids contain the carboxyl functional group –COOH -----C(=O)—OH

One of the most common polyamides is **nylon**.

Nylon is widely used as a fibre for clothing, climbing ropes and parachutes but it is also an important engineering plastic. It is a material which can be used in place of metal objects, e.g. machine parts. It has an excellent combination of strength, toughness and rigidity, and is chemically unreactive.

Nylon-6,6 is made from 2 monomers, each of which contains 6 carbon atoms.

One of the monomers is a 6-carbon di-acid, with a carboxyl acid group at each end of the molecule. The other monomer group is a di-amine, with an amino group at each end of the molecule.

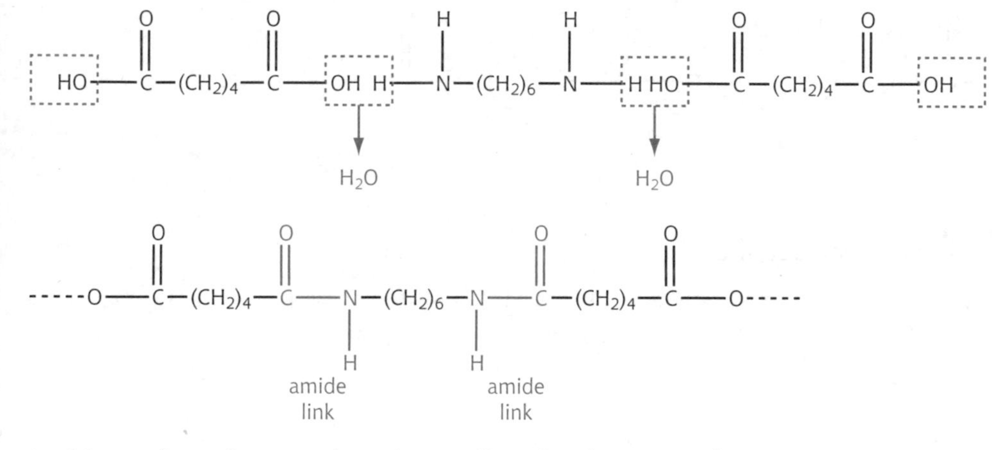

In this condensation reaction, the small molecule removed is water.

Polyamides are polymers where the repeating units are held together by amide links. An amide group has the formula CONH: -----C(=O)—N(H)-----

The amide group is formed by the reaction of the amino group with the carboxyl group.

Nylon strength

The strength of nylon comes from the hydrogen bonding between the polymer chains.

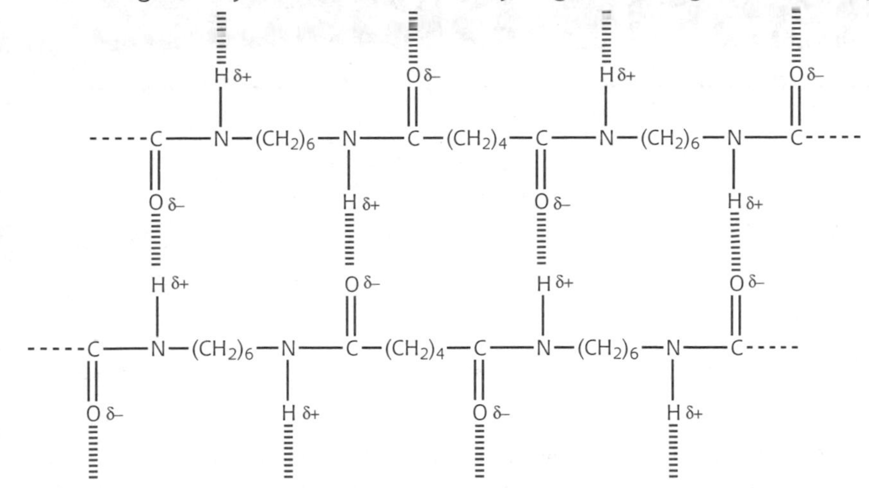

As you can see, hydrogen bonds form between the polymer chains.

This arises because of the difference in electronegativity between carbon and oxygen and between nitrogen and hydrogen.

O δ–
-----C——N-----
δ+ δ–
H δ+

Remember that hydrogen bonds are stronger than van der Waals forces. This means that polyamides have higher melting points than polyesters.

To identify the monomers, first find the amide links and split them.

Questions

1. Name the functional group in amines.
2. Draw the structure of an amide link.
3. Why is nylon strong?
4. How many monomers are present in the following section of a polyamide?

-----C(=O)—$(CH_2)_4$—C(=O)——N(H)-$CH(CH_3)$-N(H)——C(=O)—CH_2—C(=O)——N(H)—CH_2——N(H)-----

Thermosetting and thermoplastic plastics

There are 2 main types of plastic:

1. Thermoplastic

   This type of plastic can be melted and reshaped many times.

2. Thermosetting

   This type of plastic will burn or char on heating.

The structural difference between these polymers is that the thermosetting polymers have cross-links between their chains but the thermoplastic polymers do not.

When a thermoplastic polymer is heated, the chains are free to move past each other, making the sample less rigid and eventually melting it. This cannot happen with a thermosetting polymer, because its chains are locked together by the cross-links. The energy from the heat must eventually go into breaking bonds, which leads to decomposition of the polymer.

This schematic view suggests the difference:

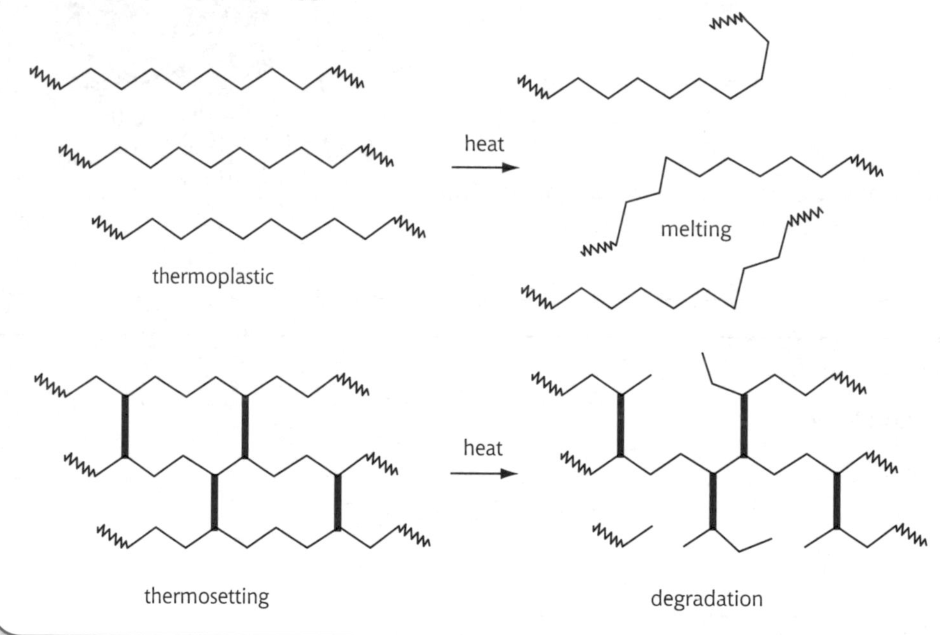

Methanal

Methanal is an important feedstock in the manufacture of thermosetting plastics. It can be made as follows:

1. Steam reforming of methane, CH_4, or coal, C, to make synthesis gas.

   Synthesis gas is a mixture of carbon monoxide, CO, and hydrogen, H_2.

   methane: $CH_4(g) + H_2O(g) \longrightarrow CO(g) + 3H_2(g)$

   coal: $C(s) + H_2O(g) \longrightarrow CO(g) + H_2(g)$

2. The synthesis gas is converted into methanol:

   $CO(g) + 2H_2(g) \longrightarrow CH_3OH(l)$

3. Methanol is oxidised to methanal:

   $CH_3OH(l) \longrightarrow HCHO(l)$

Methanal-based polymers

- Bakelite and melamine resin are 2 common examples of this type of polymer. They have cross-linked 3-D network structures, which makes them thermosetting.
- Urea-methanal

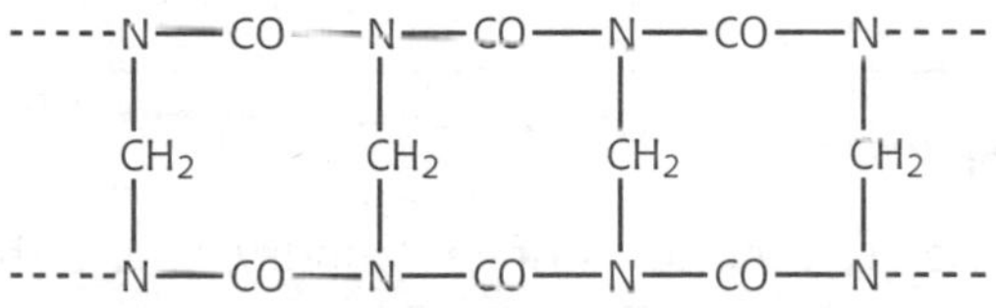

```
-----N—CO—N—CO—N—CO—N----
     |      |      |      |
    CH2    CH2    CH2    CH2
     |      |      |      |
-----N—CO—N—CO—N—CO—N----
```

The product has many cross-links, which are almost impossible to separate. The material will not melt on heating. It will eventually break down at high temperatures, decomposing and leaving a charred mass.

Questions

1. Name the process used to make methanal.
2. What type of polymers are methanal polymers?

Recent developments

Kevlar

Kevlar is an aromatic polyamide. A section of the polymer is shown:

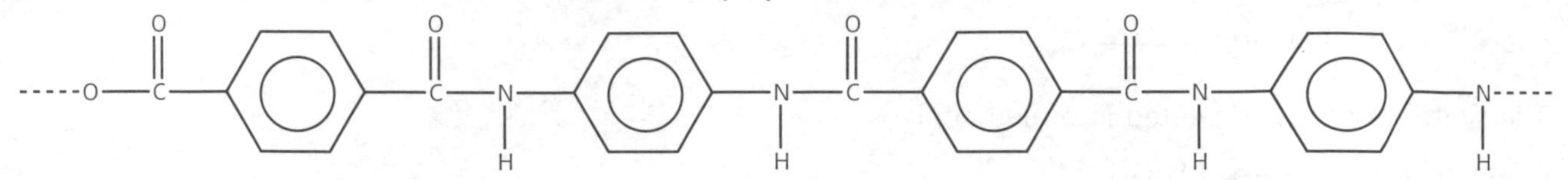

It is an extremely strong material because the rigid linear polymer chains can pack neatly on top of each other, with **hydrogen bonds between the chains**.

Kevlar's main properties are:

- it is extremely strong
- it is light
- it is fireproof

This means it can be used for making bullet-proof vests, brake pads and bicycle frames.

Poly(ethenol)

Poly(ethenol) is an unusual polymer in that it is soluble in water.

It can be used to make plastic bags which dissolve in water. These are used for hospital laundry, where the bag can be safely handled to avoid contact with contents which may be infected. The bag dissolves in the water and the laundry gets washed.

Poly(ethenol) can be made from another plastic, poly(ethenyl ethanoate), in a process called ester exchange.

Poly(ethenyl ethanoate) reacts with methanol.

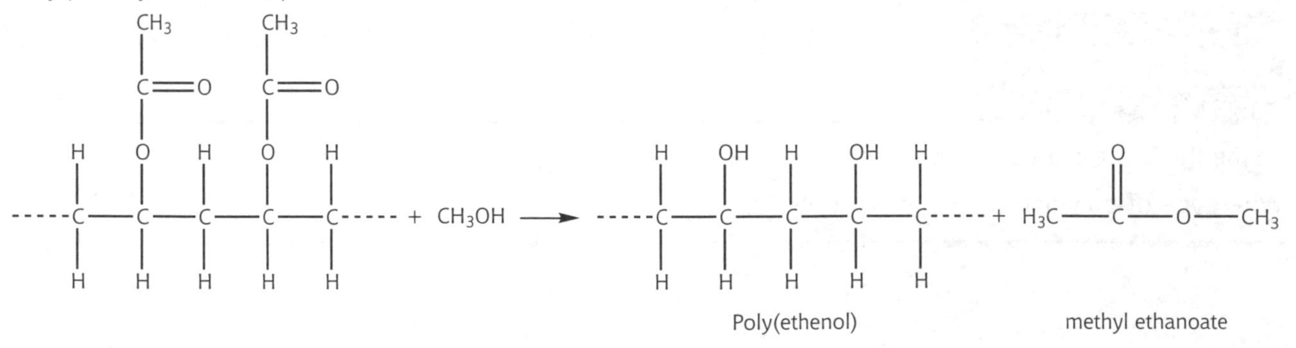

It is the hydroxyl, OH, group that makes the polymer soluble in water. The extent of the ester exchange can be controlled by warming, and the percentage of ester groups removed will determine the extent of solubility.

For example, when 90% of the ester groups have been replaced, the polymer is soluble in warm water.

Poly(ethyne)

Poly(ethyne) can be made through the addition polymerisation of ethyne molecules.

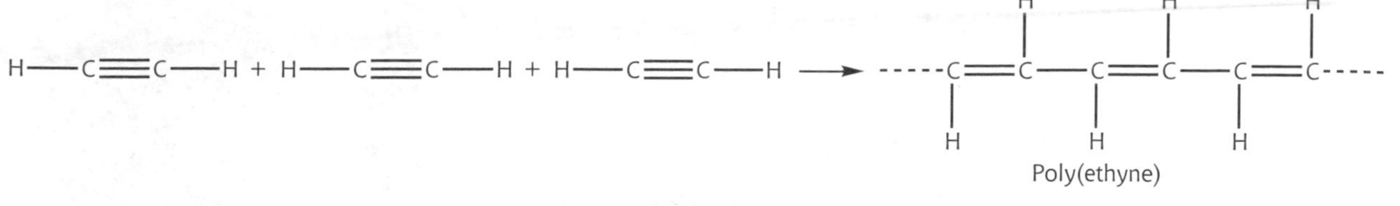
Poly(ethyne)

It is an unusual polymer in that it can conduct electricity when it has been treated with a 'dopant'. Its conductivity will depend on the delocalised electrons found along the polymer backbone.

Poly(vinyl carbazole)

Poly(vinylcarbazole) is another addition polymer made from the monomer vinyl carbazole:

It is unusual in that it exhibits photoconductivity: it will conduct electricity when light shines on it. It is this property which makes it a component in photocopiers.

Vinyl carbazole

Biopol

Biopol is a natural polyester made by certain bacteria and used by them as a source of energy.

It is a biodegradable polymer, which means it can be broken down by the action of bacteria. The advantage of biopol over a traditional plastic is that biopol can be broken down in 9 months. However, it costs about 15 times more than traditional plastics.

A section of the polymer is shown:

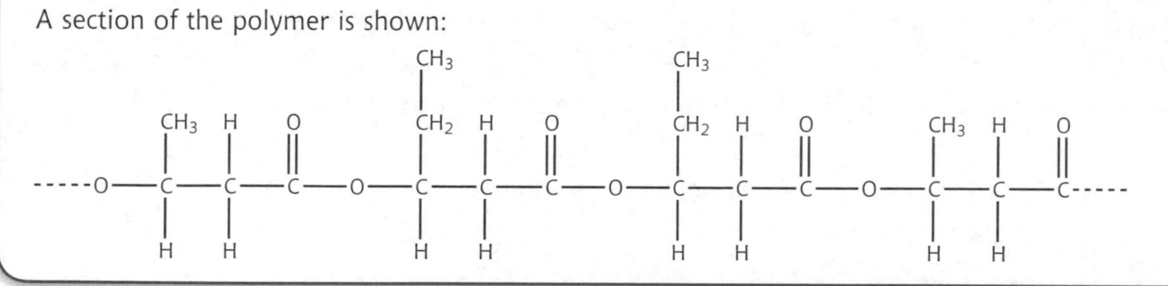

Photodegradable low-density poly(ethene)

Low-density poly(ethene) can be modified to produce a photodegradable polymer. This means it will degrade on exposure to ultraviolet light.

Questions

1. Name the polymer which:
   (*a*) dissolves in water
   (*b*) demonstrates high strength
   (*c*) is photoconductive
   (*d*) will biodegrade
   (*e*) will conduct electricity

Fats and oils

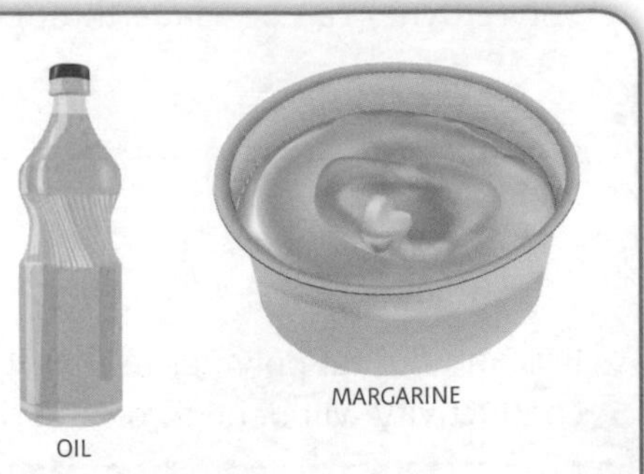

Fats and oils can be classified according to their origin:

- animal, e.g. butter fat
- vegetable, e.g. olive oil
- marine, e.g. fish oil

They supply the body with energy in a more concentrated source than carbohydrates.

Fats and oils are esters and are made through a condensation reaction between the alcohol propane-1,2,3-triol (also known as glycerol) and 'fatty acids'.

Fatty acids are saturated or unsaturated straight-chain carboxylic acids containing numbers of carbon atoms ranging from C_4 to C_{24}, but primarily between C_{16} and C_{18}.

Stearic acid is a typical saturated long-chain fatty acid: $C_{17}H_{35}COOH$:

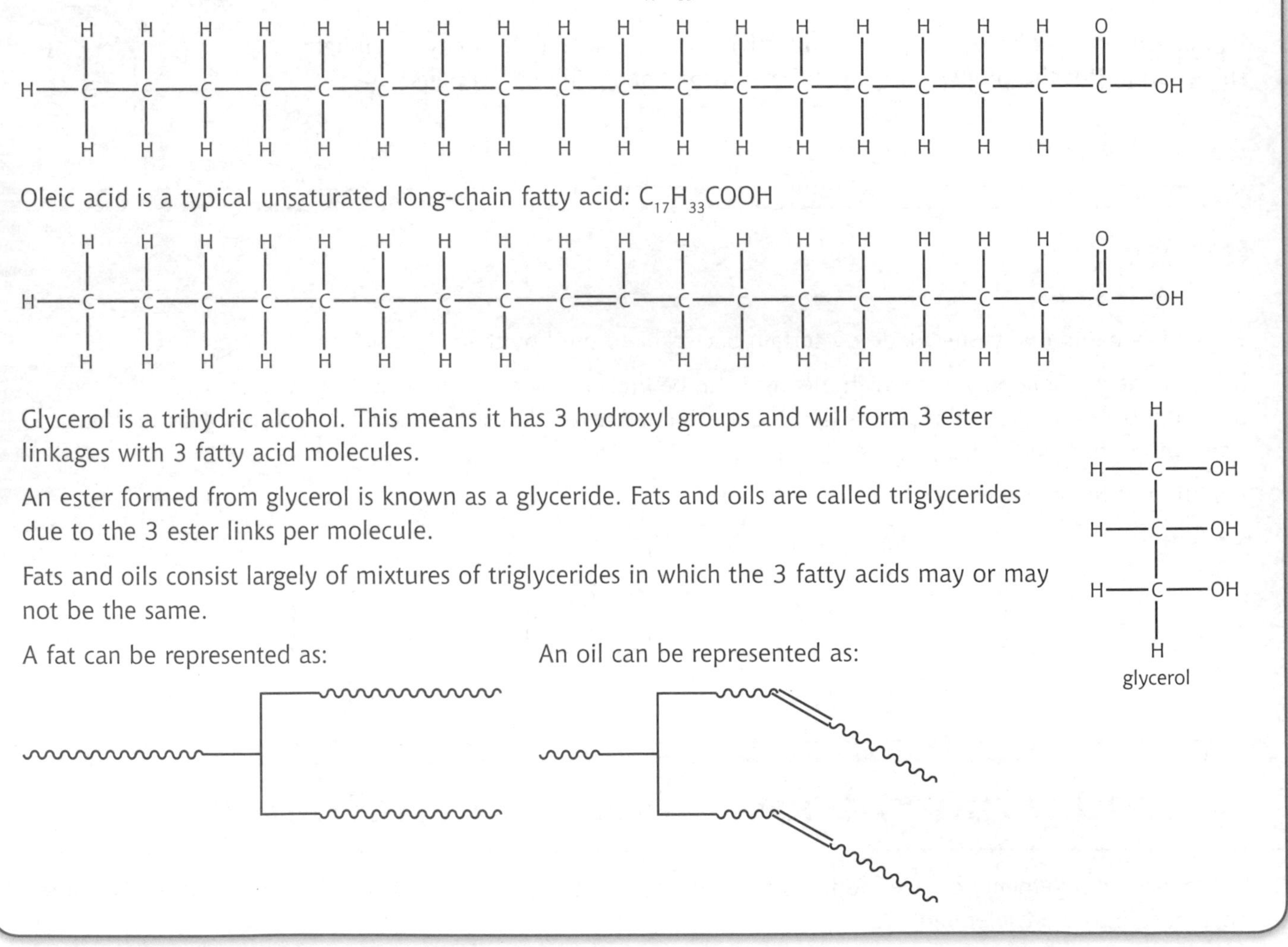

Oleic acid is a typical unsaturated long-chain fatty acid: $C_{17}H_{33}COOH$

Glycerol is a trihydric alcohol. This means it has 3 hydroxyl groups and will form 3 ester linkages with 3 fatty acid molecules.

An ester formed from glycerol is known as a glyceride. Fats and oils are called triglycerides due to the 3 ester links per molecule.

Fats and oils consist largely of mixtures of triglycerides in which the 3 fatty acids may or may not be the same.

A fat can be represented as:

An oil can be represented as:

Structures of fats and oils

Fats have higher melting points than oils, which means they are generally in solid state at room temperature. This is due to the way the molecules pack together. A fat is made up of saturated fatty acid chains and this allows for the closer packing.

Fat

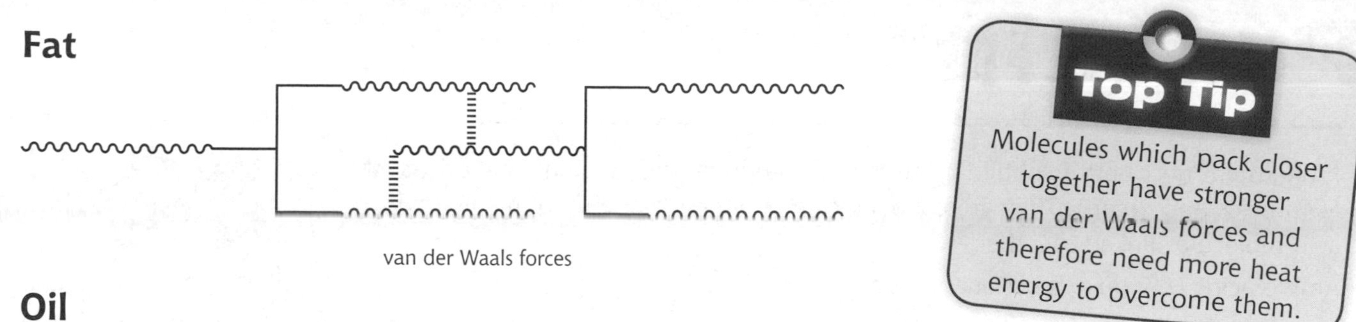

Top Tip

Molecules which pack closer together have stronger van der Waals forces and therefore need more heat energy to overcome them.

Oil

Oils are made up of unsaturated fatty acid chains and this considerably alters how closely the triglyceride molecules can pack together. The C=C bond distorts the chain and prevents the oil molecules from packing closely together.

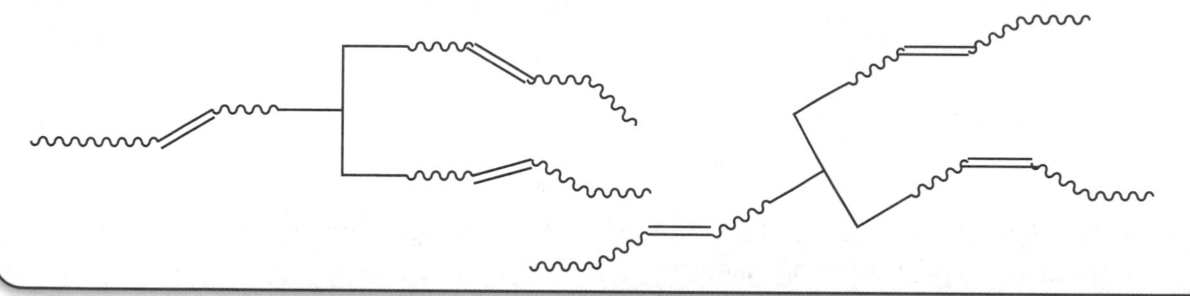

Reactions of fats and oils

Oils can be converted into fats by the addition of hydrogen across the carbon to carbon double bonds in the unsaturated fatty acids. As the liquid oil is turned into a solid, the process can be described as **hardening**.

This reaction is carried out using a nickel catalyst, and since hydrogen is added it is also called **hydrogenation**.

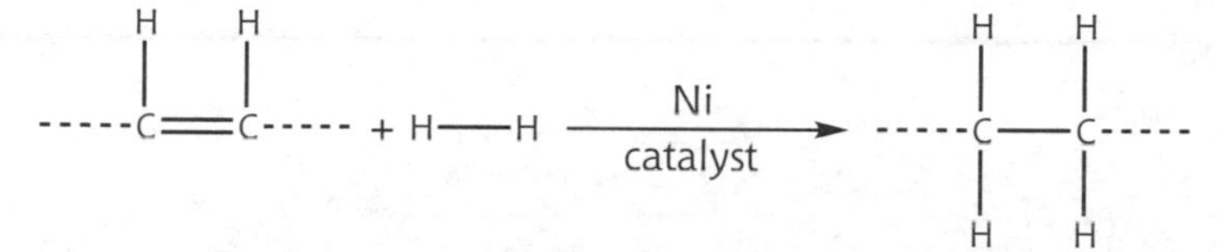

Hydrolysis of fats and oils

As with all esters, fats and oils can be broken down or hydrolysed into their alcohol and acid. With fats and oils, the reaction will produce 1 mole of glycerol to 3 moles of fatty acids.

Soaps are produced by the hydrolysis of fats and oils – see page 89.

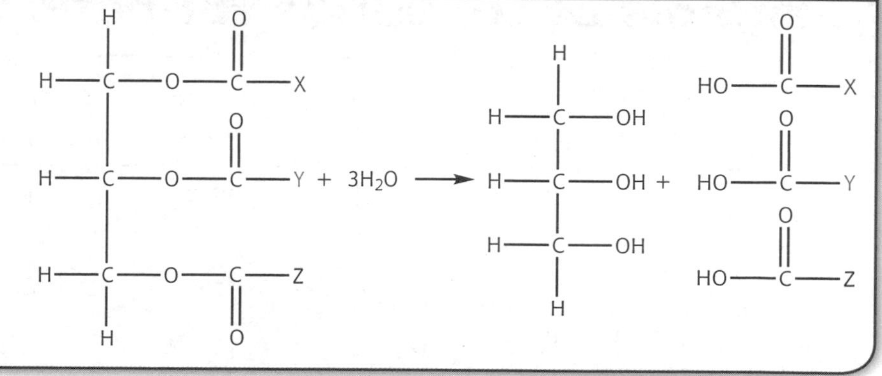

Questions

1. Why are oils liquid?
2. Name the alcohol used to make fats and oils.
3. How many ester linkages are there in a fat?
4. Name the process used to convert oil into a fat.

Proteins

Proteins are necessary for building structural components of the human body, such as muscles and organs. They are condensation polymers made up of many amino acid molecules linked together.

Amino acids contain 2 functional groups:

```
                        O                                  H
                        ||                                 |
Carboxyl group   -----C----OH        Amino group    -----N----H
```

Remember: nitrogen is essential for protein formation by plants and animals.

There are 20 different amino acids which are used to make the proteins needed by the body.

The structure of a typical amino acid is shown:

```
       O    H    H
       ||   |    |
  HO---C----C----N----H
            |
            R
```

The **R** group can contain different elements – carbon, hydrogen, sulphur, nitrogen and oxygen. It is this **R** group which will differentiate 1 amino acid from another.

When **R = H,** the amino acid is called glycine:

When **R = CH_3,** the amino acid is called alanine:

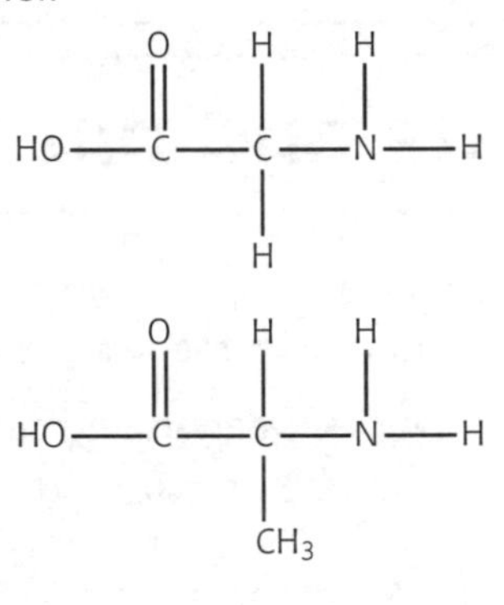

Amino acids join together in a condensation reaction:

- 2 amino acids join to make a dipeptide
- 3 amino acids join to make a tripeptide
- many amino acids join to make a polypeptide, also called a protein.

Example of tripeptide formation

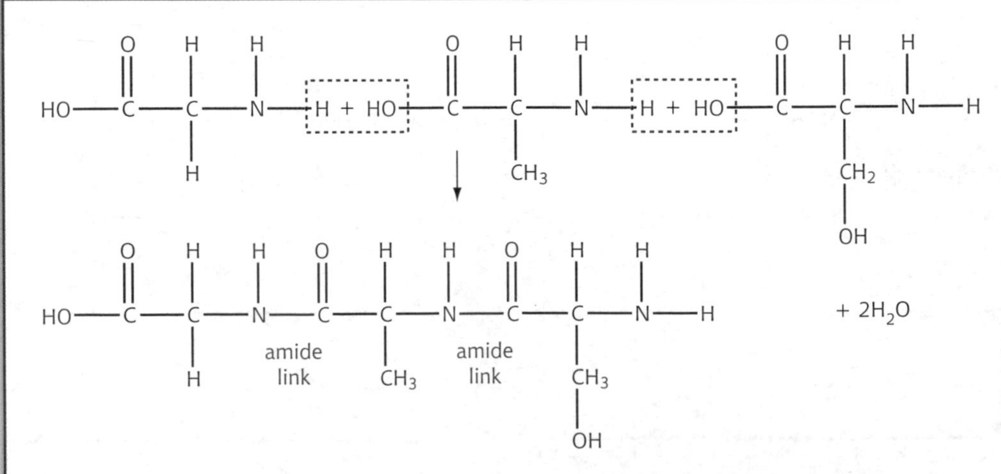

The peptide (amide) link is formed by the reaction between the carboxyl group and the amino group.

The human body can produce most of the amino acids needed to make the proteins required by the body. However, some amino acids cannot be made by the body and must come from food. **These are the essential amino acids**.

Digestion

During digestion, the proteins are hydrolysed; i.e. get broken down into their amino acids. Water molecules break the peptide link.

During digestion, the animal and vegetable proteins are hydrolysed and the amino acids pass into the bloodstream. The proteins required by the body are built up from the amino acids in the bloodstream.

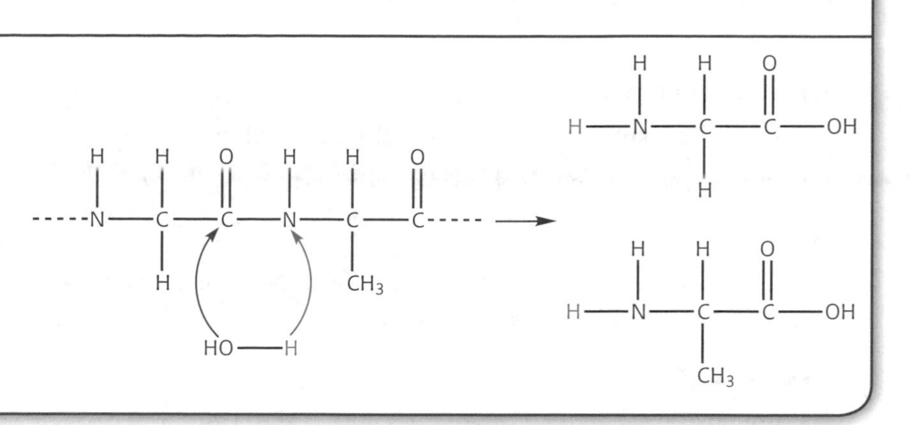

Types of protein

There are 2 types of proteins, fibrous and globular

Fibrous proteins are long and thin and are the major structural materials of animal tissues. Examples are keratin, found in hair, and collagen, found in tendons.

Globular proteins are involved in the maintenance and regulation of life processes and have spiral chains folded into a compact shape.

Examples are enzymes (e.g. ribonuclease), hormones (e.g. insulin) and haemoglobin.

Enzymes

All enzymes are biological catalysts. Each enzyme catalyses a specific reaction, which is shown in the fermentation reaction of starch to ethanol:

starch $\xrightarrow{\text{amylase}}$ maltose $\xrightarrow{\text{maltase}}$ glucose $\xrightarrow{\text{zymase}}$ ethanol

It is believed that enzymes work on a 'lock and key' principle.

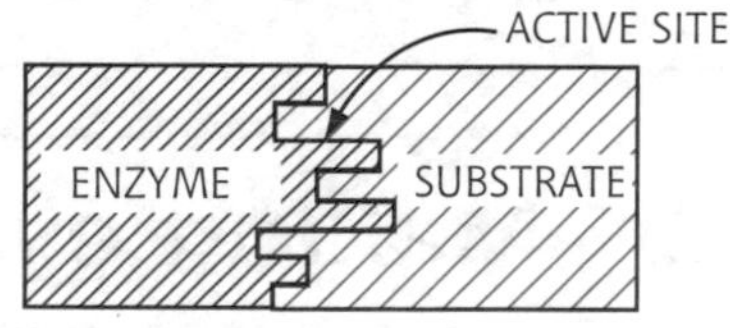

The specific shape of the active site on enzymes only allows certain substrate molecules to fit into it.

Enzymes are specific, that is they work best in narrow temperature and pH ranges. Outwith these ranges, enzymes become denatured. This means that they lose their specific shapes and their substrate molecules can no longer fit inside. The reactions slow down and eventually stop.

Questions

1. What type of molecule are all proteins made up of?
2. Name the link found in proteins.
3. What type of protein is found in hair?

PPA 1 – Oxidation

Introduction

Aldehydes and ketones both contain the carbonyl group.

Aldehydes will oxidise to carboxylic acids, ketones will not.

Aim

To distinguish between an aldehyde and a ketone using oxidising agents – acidified potassium dichromate, Fehling's solution and Tollens' reagent.

Procedure

The unknown carbonyl compounds X and Y were heated in a water bath with the oxidising agents acidified potassium dichromate, Fehling's reagent and Tollens' reagent.

x+
Oxidising agent
y+
Oxidising agent
HEAT

Results

The following results were obtained:

Oxidising agent	Observation with X	Observation with Y
acidified potassium dichromate	turns from orange to blue-green	stays orange
Fehling's solution	turns from blue to brick-red	stays blue
Tollens' reagent	turns from colourless to a silver mirror appearing on the sides of the test-tube	no silver mirror forms

Conclusion

Oxidising agents can be used to distinguish between aldehydes and ketones. The aldehyde gave positive results and the ketone did not, therefore X is the aldehyde.

Evaluation

A water bath is used to heat the solutions, because carbonyl compounds are highly flammable.

PPA 2 – Making esters

Introduction

An ester is made by reacting an alcohol with a carboxylic acid. Concentrated sulphuric acid is used as a catalyst in the experiment.

Aim

To prepare an ester and to identify some of its characteristics.

Procedure

Equal volumes of an alcohol (methanol) and an alkanoic acid (butanoic acid) were added to a boiling tube. A few drops of concentrated sulphuric acid were then added. A wet paper towel was secured around the outside of the boiling tube with an elastic band. A cotton-wool plug was inserted into the mouth of the boiling tube, which was then placed in a hot water bath for 15 minutes. After this time the contents of the boiling tube were poured into a beaker containing sodium hydrogen carbonate solution. This reaction gave off vigorous effervescence.

Results

The ester floated on top of the aqueous solution and had a pleasant smell, of pineapple drops.

Conclusion

This procedure made the ester methyl butanoate.

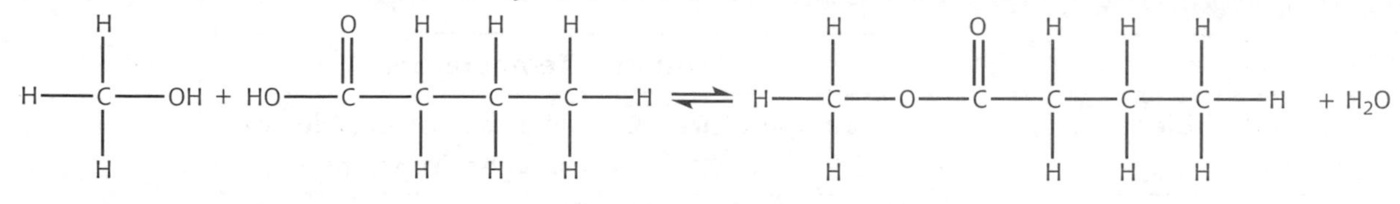

Esters will float on water. They do not mix with water, giving the characteristic 2-layer effect, and they have a 'pleasant' smell.

Evaluation

- The procedure is carried out using a water bath, because alcohols and carboxylic acids are highly flammable.
- The purpose of the wet paper towel is to act as a condenser and to prevent gases from escaping.
- The reaction is speeded up by using a catalyst and by heating.
- The purpose of the cotton wool is to absorb any reactants which may spurt out of the boiling tube during heating.
- The purpose of the sodium hydrogen carbonate is to neutralise the concentrated sulphuric acid and any of the unreacted carboxylic acid.

PPA 3 – Factors affecting enzyme activity

Introduction

Hydrogen peroxide will slowly decompose to form water and oxygen at room temperature. The enzyme catalase will speed up this reaction: $2H_2O_2 \longrightarrow 2H_2O + O_2$

Aim

To investigate the effect of pH or temperature changes on enzyme activity.

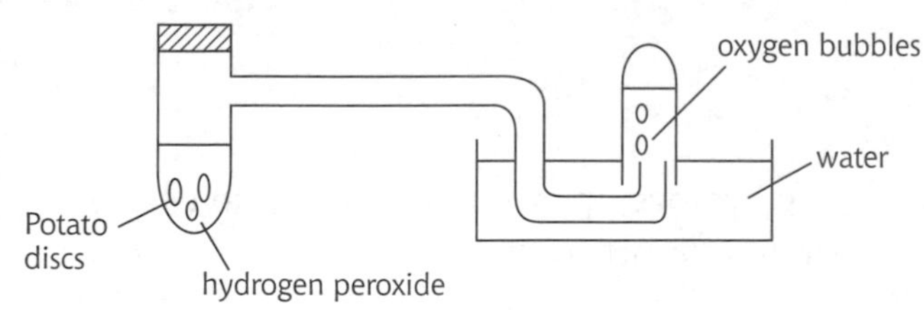

Procedure – pH

3 potato discs were added to a side-arm tube and 5 cm³ of pH7 buffer solution was added. The mixture was allowed to stand for 5 minutes. 1 cm³ of hydrogen peroxide solution was then added and the number of bubbles of gas given off in 3 minutes was recorded. The experiment was repeated using pH buffer solutions of 1, 4, 10 and 13.

Procedure – temperature

3 potato discs were added to a side-arm tube and 5 cm^3 of water was then added. The side-arm tube was placed in a water bath at 20°C. It remained there for 5 minutes so the temperature of the mixture became constant. 1 cm^3 of hydrogen peroxide solution was added and the number of bubbles of oxygen released in 3 minutes was counted. The experiment was repeated at 30°C, 40°C, 50°C and 60°C.

Results – pH	
pH	Number of bubbles of oxygen in 3 mins
1	2
4	8
7	18
10	13
13	3

Results – Temperature	
Temperature °C	Number of bubbles of oxygen in 3 mins
21	9
29	14
42	31
51	6
61	2

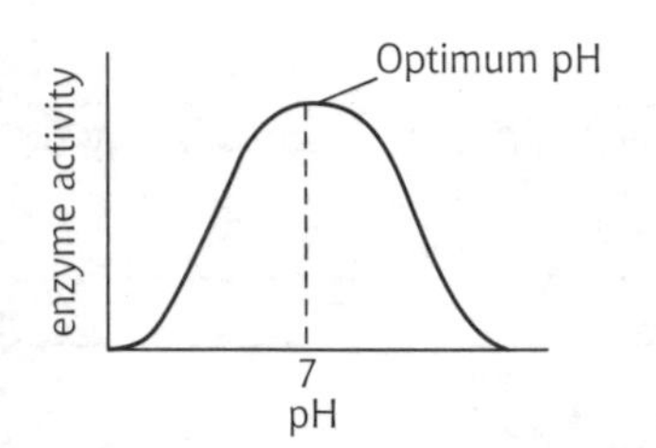

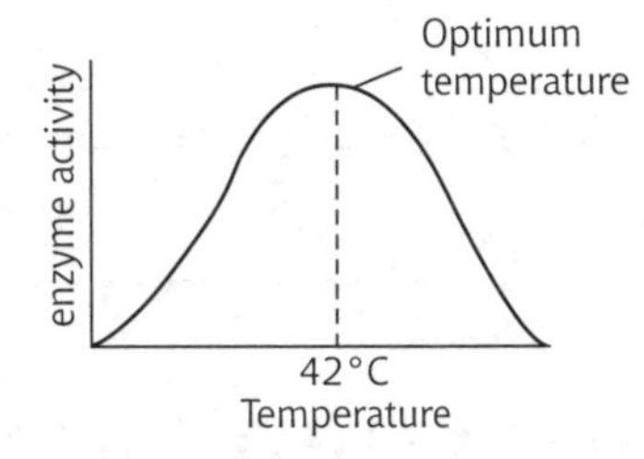

Conclusion – pH

The enzyme catalase has an optimum activity at pH7. Activity decreases in strong acid or alkaline conditions.

Conclusion – Temperature

The enzyme catalase has an optimum activity at 42°C. Activity decreases at temperatures lower and higher than 42°C.

Evalution

- Both experiments were left to stand for 5 minutes before the hydrogen peroxide was added to ensure that the enzyme, catalase, had time to adjust to the temperature or pH conditions.
- When investigating pH, only the pH solution was changed. Other factors such as temperature, number of potato discs and volume of buffer solution were kept the same.
- When investigating temperature, only the temperature was changed. All other factors remained the same.

The World of Carbon

Summary of organic compounds

Name	Functional group
alkene	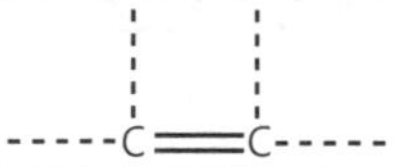 C to C double bond
alkyne	 -----C≡C----- C to C triple bond
halogenalkane	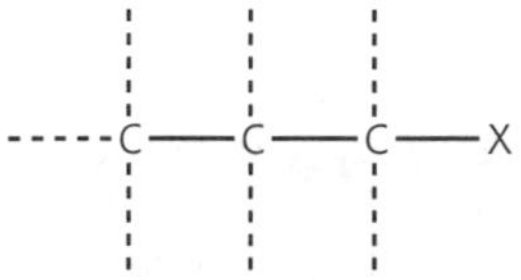 -----C—C—C—X X = F, Cl, Br, I
aldehyde or alkanal	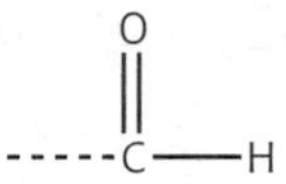 carbonyl group
ketone or alkanone	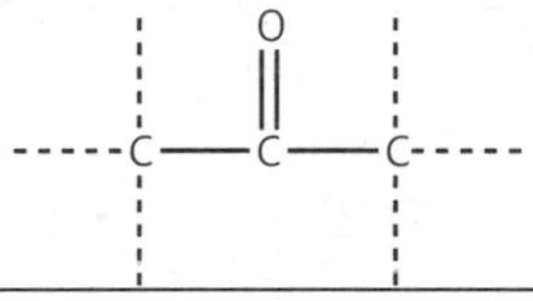 carbonyl group
alkanol or alcohol	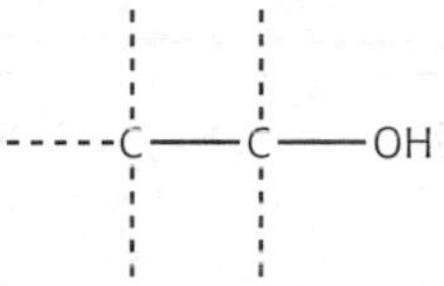 -----C—C—OH hydroxyl group
alkanoic acid or carboxylic acid	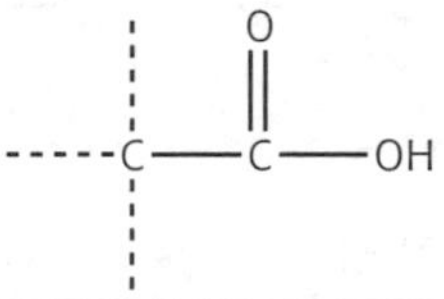 carboxyl group
ester	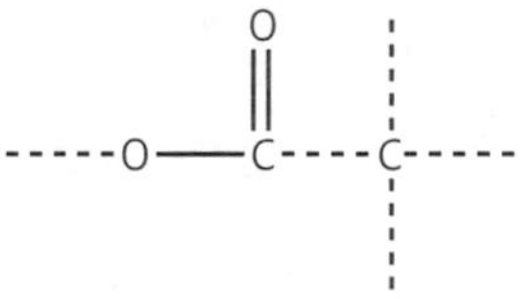
aromatic	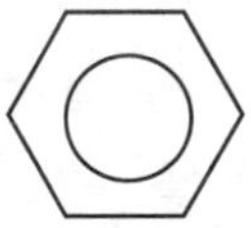 benzene or phenyl group
amino acid	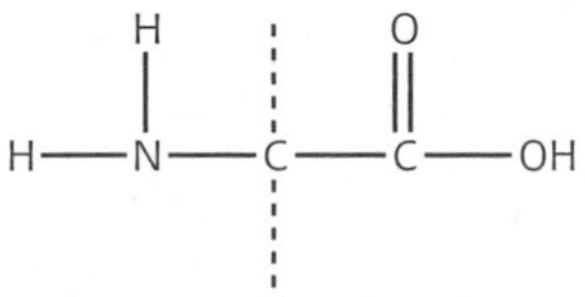

Summary of organic reactions

Type of reaction	Example
addition	alkene ⟶ halogen alkane $-C{=}C- + Br_2 \longrightarrow -C(Br)-C(Br)-$
hydration	alkene ⟶ alcohol $-C{=}C- + H_2O \longrightarrow -C-C-OH$
dehydration	alcohol ⟶ alkene $-C-C-OH \longrightarrow -C{=}C- + H_2O$
combustion	burning in oxygen $C + O_2 \longrightarrow CO_2$
oxidation	primary alcohol ⟶ aldehyde ⟶ carboxylic acid $-C-CH_2-OH \longrightarrow -C-C(=O)-H \longrightarrow -C-C(=O)-O-H$ secondary alcohol ⟶ ketone $-C-CH(OH)-C- \longrightarrow -C-C(=O)-C-$

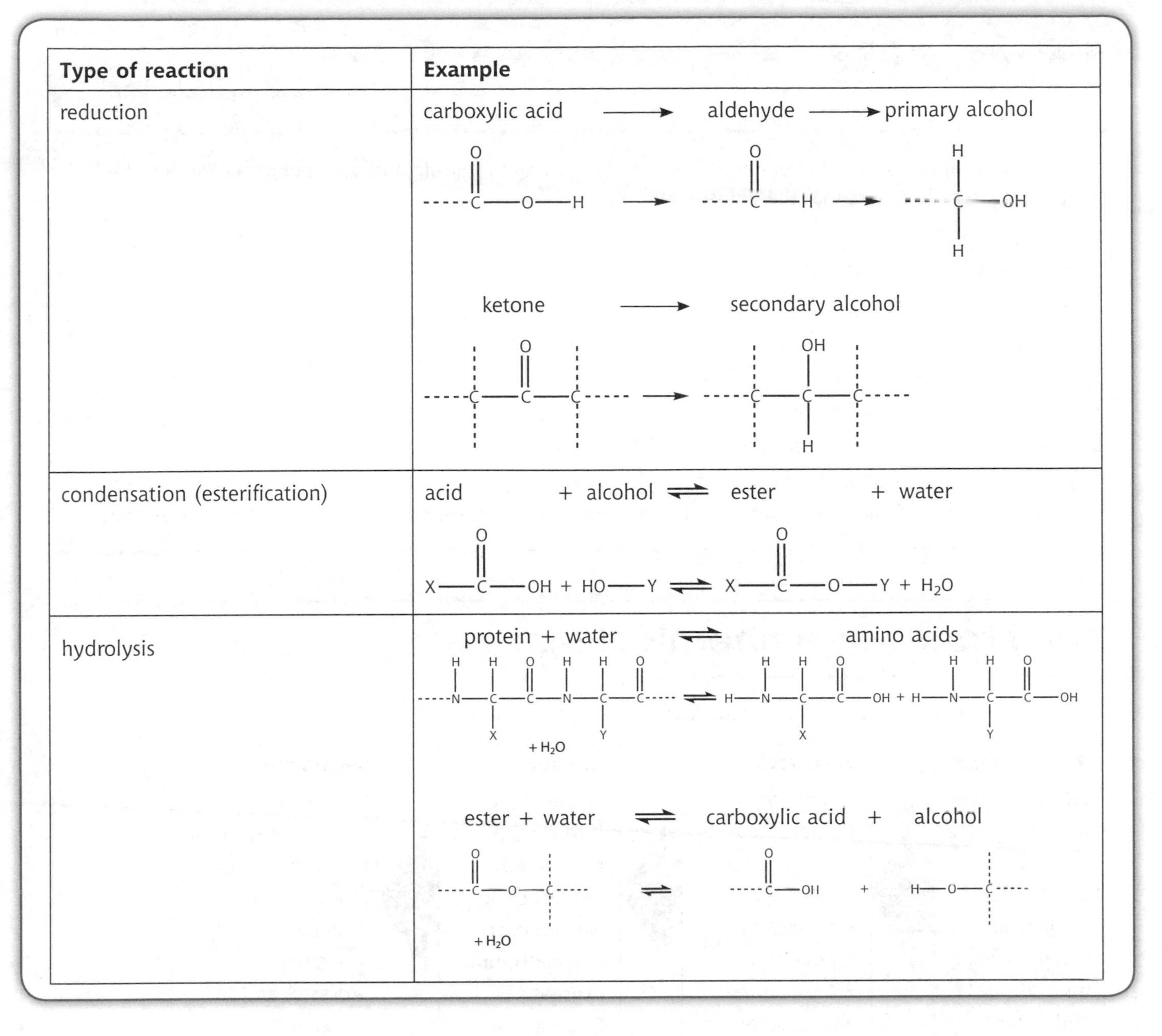

Type of reaction	Example
reduction	carboxylic acid → aldehyde → primary alcohol ketone → secondary alcohol
condensation (esterification)	acid + alcohol ⇌ ester + water $X{-}CO{-}OH + HO{-}Y \rightleftharpoons X{-}CO{-}O{-}Y + H_2O$
hydrolysis	protein + water ⇌ amino acids ester + water ⇌ carboxylic acid + alcohol

The UK chemical industry

The UK chemical industry is a major contributor to the quality of our life and our national economy. The five main categories of product that the industry makes are:

- pharmaceuticals
- petrochemicals and polymers
- paints and pigments
- speciality chemicals
- inorganics and fertilisers

The largest sector is pharmaceuticals, with 37% of the market. The chemical industry accounts for 14% of the UK manufacturing industry, employing around 200,000 people in 3,100 companies.

The industry spends nearly £2 billion annually on new capital investment (new plants) and research and development. The UK chemical industry is, by and large, **capital intensive** rather than **labour intensive**.

The stages in a chemical plant

A chemical manufacturing process usually involves a sequence of steps:

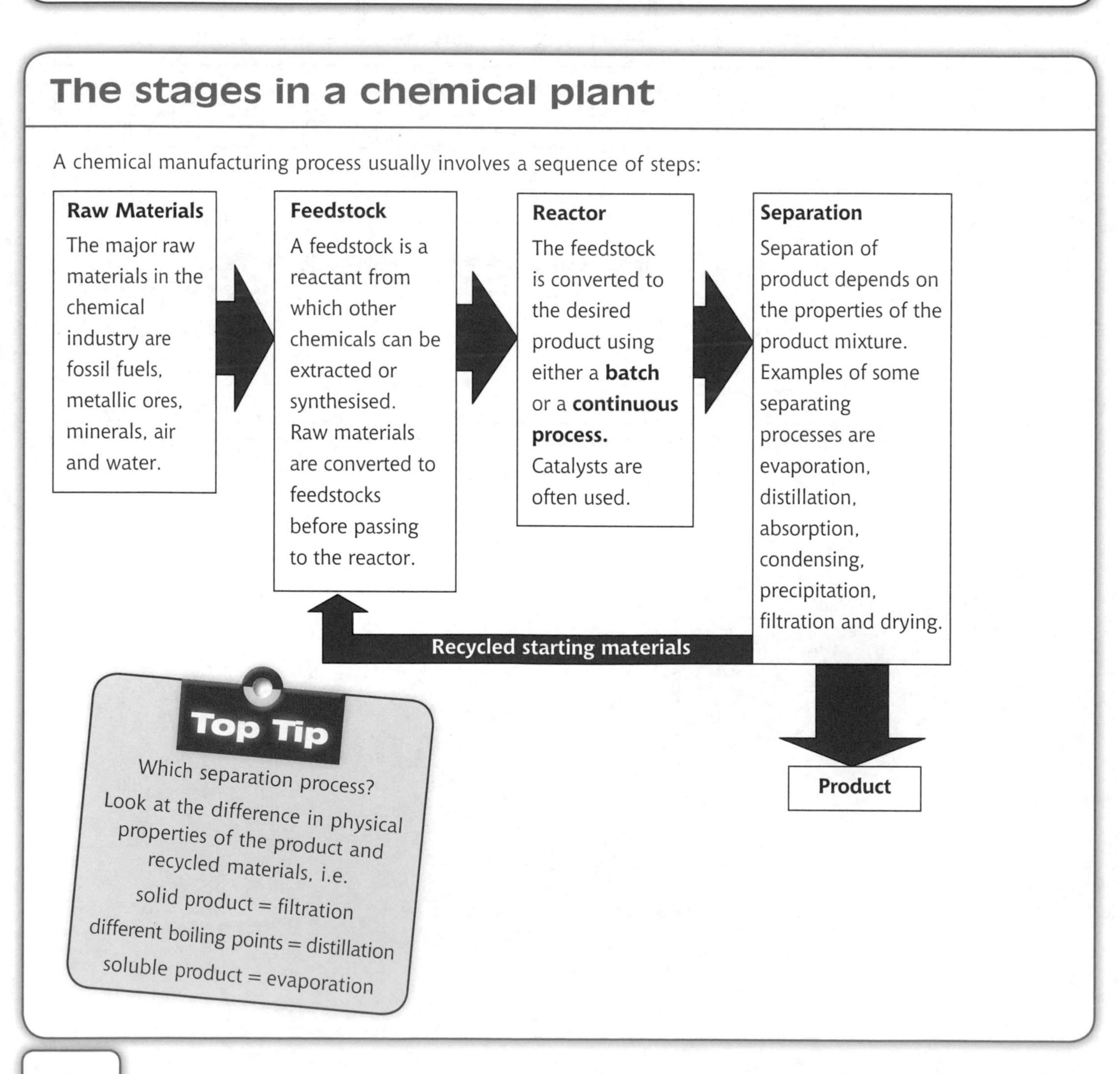

Important factors

The following factors are considered when choosing a manufacturing route. Attention to the economics of each stage is important because the main aim of the manufacturer is to obtain profit.

Flow diagrams – look for possible recycling routes for unreacted starting materials.

- Cost and availability of feedstock
- Yield of product
  Because the main consideration is the profit margin, sometimes a compromise is made between yield, catalyst and operating procedures. An example of this is the Haber process.
- Can unreacted starting materials be recycled?
  Unreacted nitrogen and hydrogen are recycled in the Haber process.
- Can by-products be sold?
  Slag, which is a by-product of iron manufacturing, is sold as a building material.
- Difficulty and cost of waste disposal.
- Energy consumption.
  Some chemical plants utilise heat given off by reactions to heat water, to produce steam, which in turn generates electricity. Production of sulphuric acid is an example.

Other factors affecting the location of a chemical plant

Historical and practical factors

Many plants are located near main transport routes such as motorways and ports, reducing the cost and increasing the ease of obtaining raw materials and transporting products. Chemical plants are often located near existing industries, allowing access to feedstocks and a skilled labour force.

Safety and environmental issues

The safety of the work force and people in the surrounding areas is of extreme importance. Plant design is constantly improving to protect workers, and legislation exists to protect workers in their work environment.

Release of by-products into the atmosphere or water system has reduced greatly. Most unreacted materials are recycled, and by-products are recovered and reused rather than removed.

Types of process

Chemical manufacturing processes can be classified as either batch or continuous processes.

Batch processes are used when small quantities of chemicals are required.
Products produced by this process are pharmaceuticals, dye-stuffs and pesticides.

Continuous processes are generally used to produce single products. They can operate 365 days a year. Products produced by such processes are mainly in the petrochemicals sector, but ammonia and sulphuric acid are other well-known examples.

Process	Advantage	Disadvantage
Batch	• Can produce a variety of products. • Plants with batch reactors are less expensive to build. • Can be used for slow reactions. • Reactants can be in any state.	• Very labour intensive. • Lost production when filling and emptying reactors. • Lost production when changing from one product to another because reactors often need to be cleaned.
Continuous	• Less labour intensive: some of the processes are automated. • Quality control of the product is more easily ensured. • Continuous operation makes for economic efficiency, because shutdowns may be months or even years apart.	• Normally designed for a specific feedstock. • Plants with continuous reactors are very expensive to build. • Most solid reactants need to be fine powders, because larger particles block pipes.

Stages in the manufacture of new products

Stages in the manufacture of a new product vary from those in an established product.

For a completely new product, the following steps are taken.

Stage	Definition
Research and development	Identification of the new product and development of a suitable manufacturing process.
Pilot study	The process is then evaluated in a pilot plant, i.e. a scaled-down version of the possible full-size plant. The pilot study uses the route identified in the research stage. Product quality, health hazards and production costs are evaluated.
Scaling up	Using information from the pilot study, the process is scaled up from kilogram quantities to full-production quantities. This involves the planning and development of a full-size plant.
Production	New product is manufactured after start-up of production.
Review	At all stages in the manufacture of a new product, the processes are reviewed and modifications are made. Particular attention is drawn to reducing costs and hazards to health and safety and environmental hazards.

Manufacturing costs

Three types of cost are incurred in operating a chemical plant:

Cost	Definition
Capital	The initial cost of building the plant, research and development and associated infrastructure.
Fixed	Fixed costs are costs which are incurred whether a plant is operating at maximum capacity or only at partial capacity. • Salaries • Depreciation (maintenance of plant) • Sale expenses
Variable	Variable costs are determined by plant output. • Total cost of raw materials • Total cost of distribution of product • Total energy costs • Waste product treatment or disposal

Questions

1. Aspirin is produced by a batch process. What are the advantages of a batch process over a continuous process?

2. Which of the following is an example of a fixed cost?

(*a*) maintenance of plant

(*b*) cost of raw materials

(*c*) cost of energy

(*d*) research and development

3. Which of the following is a raw material?

(*a*) naphtha

(*b*) sodium chloride

(*c*) ammonia

(*d*) sulphur dioxide

4. Name the five stages in the manufacture of a new product.

Hess's law

Enthalpy change

Unit One (pages 14–16) explains the method used to determine the enthalpy of combustion, solution and neutralisation using experimental data.

The enthalpy change for a number of reactions cannot be determined in this way and so an alternative route, using Hess's Law, is taken.

Hess's law states that the **enthalpy change for a chemical reaction is independent of the route taken**.

PPA 1 – Hess's law

Introduction

Potassium hydroxide can be converted into potassium chloride by two alternative routes.

The direct route

Solid potassium hydroxide reacts with hydrochloric acid.

The indirect route

A solution of potassium hydroxide is made by dissolving solid potassium hydroxide in water. The resulting solution is then reacted with hydrochloric acid.

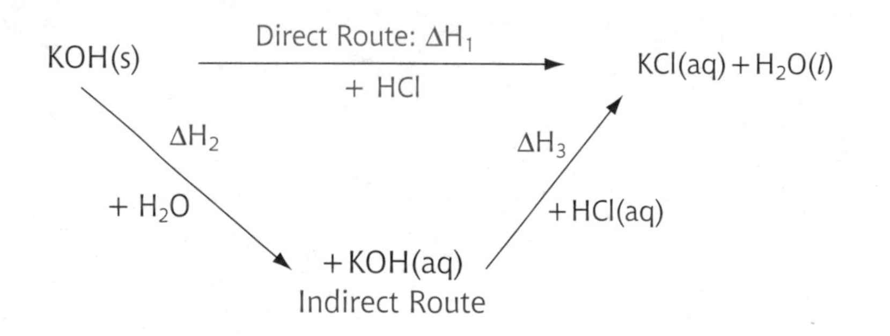

Hess's law states that the enthalpy change will be the same no matter whether the direct or indirect route is taken.

The aim of this PPA is to confirm Hess's law by following the formation of potassium chloride solution by direct and indirect routes.

Direct route: $KOH(s) + HCl(aq) \longrightarrow KCl(aq) + H_2O(l)$

Procedure

- Potassium hydroxide was accurately weighed (no more than 1.4 g) and placed in a polystyrene cup.
- 25 cm^3 of 1 mol l^{-1} HCl was measured out using a measuring cylinder, and the temperature measured and recorded.
- The acid was added to the potassium hydroxide and stirred. The highest temperature reached was then recorded.

Indirect route:

$KOH(s) + H_2O(l) \xrightarrow{\text{Step 1}} KOH(aq)$

$KOH(aq) + HCl(aq) \xrightarrow{\text{Step 2}} KCl(aq) + H_2O(l)$

Step 1

- 25 cm^3 of water was measured out and its temperature recorded.
- Potassium hydroxide was weighed out (no more than 1.4 g) and placed in a polystyrene cup.
- The water was added to the potassium hydroxide and stirred; the highest temperature reached was recorded. This solution was kept for Step 2.

Step 2

- 25 cm^3 of 1 mol l^{-1} HCl was measured out using a measuring cylinder and placed in a polystyrene cup.
- The temperature of the acid was measured and recorded.
- The temperature of the potassium hydroxide solution from Step 1 was measured and recorded.
- The alkali was added to the acid and stirred, and the highest temperature was measured and recorded.

The initial temperature is taken as an average of the acid and alkali.

Conclusion

The enthalpy of the direct route is the same as that for the sum of the indirect route. Hess's law was proved.

Evaluation

- Some heat will be lost to the surroundings, such as the container or the air. This means that not all the heat will be transferred.
- Using polystyrene instead of glass beakers will reduce heat loss.

Questions

1. List the measurements that would have to be taken in order to calculate the indirect route.
2. The experimental enthalpy change was less than that published due to heat loss to surroundings. What can be done to lower the heat loss?

Applying Hess's law

Calculate the enthalpy change (ΔH) for the formation of 1 mole of benzene

Step 1 – Write the forward reaction – this is the target equation:

$6C(s) + 3H_2(g) \longrightarrow C_6H_6(l)$

Top Tip

When balancing equations leave oxygen to last.

Step 2 – Write the standard enthalpy of combustion equation for each reactant and product in the equation. See page 14 in Unit 1.

Write the standard enthalpy value next to the equation. (See page 6 in your data booklet.)

$C(s) + O_2(g) \longrightarrow CO_2(g) \quad \Delta H = -394\,kJ\,mol^{-1}$

$H_2(g) + 1/2O_2(g) \longrightarrow H_2O(l) \quad \Delta H = -286\,kJ\,mol^{-1}$

$C_6H_6(g) + 7 1/2O_2(g) \longrightarrow 6CO_2(g) + 3H_2O(l) \quad \Delta H = -3268\,kJ\,mol^{-1}$

Step 3 – The standard enthalpy values in the data booklet are for one mole of a substance burning. Therefore you must multiply this value to build up the required equation.

Six moles of carbon are required: $(-394 \times 6)\,kJ\,mol^{-1}$

$6C(s) + 6O_2(g) \longrightarrow 6CO_2(g) \quad \Delta H = (-394 \times 6)\,kJ\,mol^{-1} = -2364\,kJ\,mol^{-1}$

Three moles of hydrogen are required: $(-286 \times 3)\,kJ\,mol^{-1}$

$3H_2(g) + 1 1/2O_2(g) \longrightarrow 3H_2O(l) \quad \Delta H = (-286 \times 3)\,kJ\,mol^{-1} = -858\,kJ\,mol^{-1}$

Step 4 – Equations may need to be reversed to build up the desired calculation.

$C_6H_6(g) + 7 1/2O_2(g) \longrightarrow 6CO_2(g) + 3H_2O(l) \quad \Delta H = -3268\,kJ\,mol^{-1}$

When you reverse the equation you reverse the sign.

$6CO_2(g) + 3H_2O(l) \longrightarrow C_6H_6(g) + 7 1/2O_2(g) \quad \Delta H = +3268\,kJ\,mol^{-1}$

Step 5 – Add up all the equations to get the desired equation. Score out any species which occur as both products and reactants. You should be left with the target equation.

$6C(s) + \cancel{6O_2(g)} \longrightarrow \cancel{6CO_2(g)} \quad \Delta H = -2364\,kJ\,mol^{-1}$

$3H_2(g) + \cancel{1 1/2O_2(g)} \longrightarrow \cancel{3H_2O(l)} \quad \Delta H = -858\,kJ\,mol^{-1}$

$\cancel{6CO_2(g)} + \cancel{3H_2O(l)} \longrightarrow C_6H_6(l) + \cancel{7 1/2O_2(g)} \quad \Delta H = +3268\,kJ\,mol^{-1}$

$6C(s) + 3H_2(g) \longrightarrow C_6H_6(l)$

Step 6 – Add all the enthalpy values to obtain the final answer:

$\Delta H = (-2364) + (-858) + 3268$

$= +46\,kJ\,mol^{-1}$

An alternative approach to using Hess's law

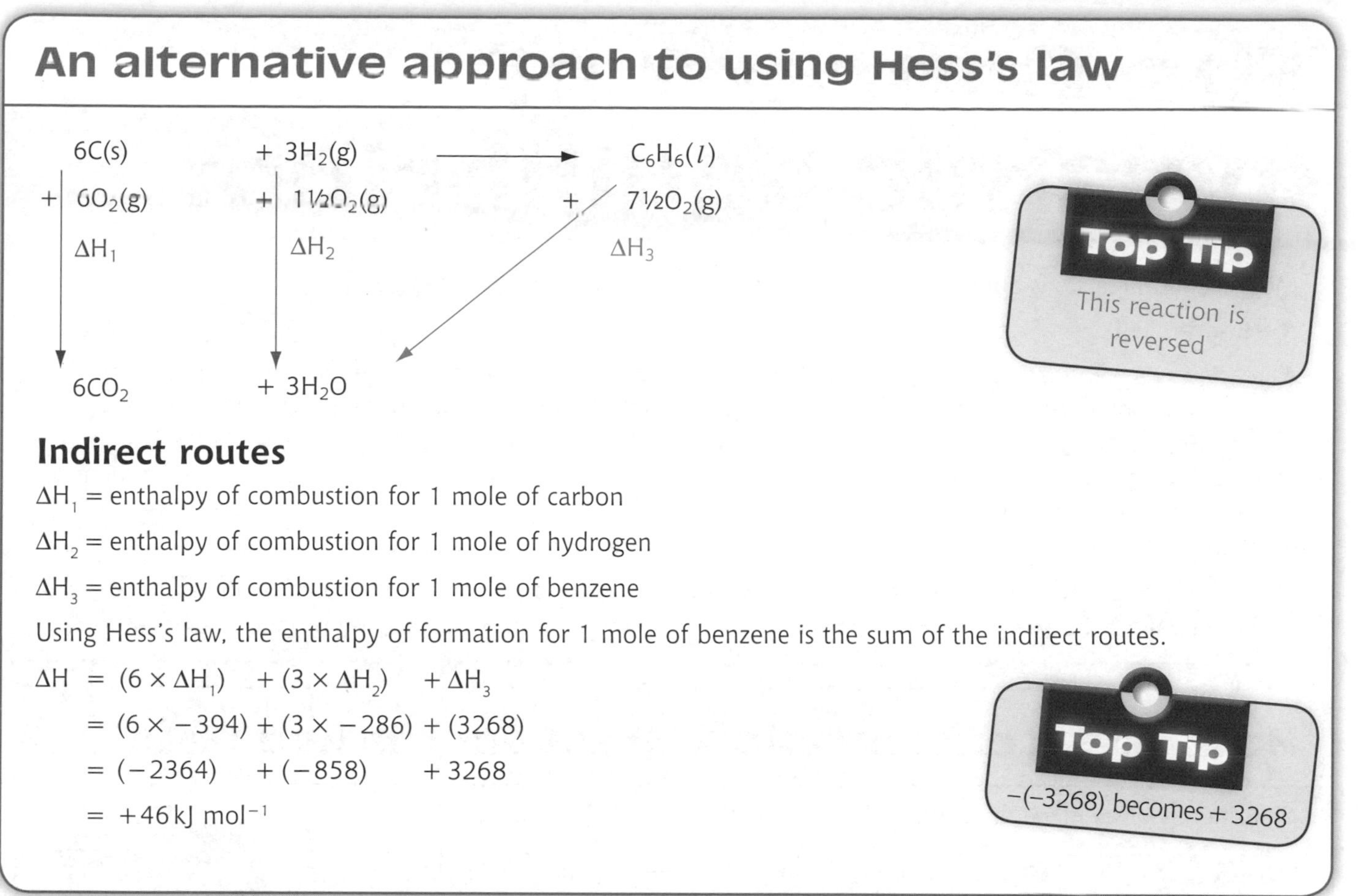

Indirect routes

ΔH_1 = enthalpy of combustion for 1 mole of carbon

ΔH_2 = enthalpy of combustion for 1 mole of hydrogen

ΔH_3 = enthalpy of combustion for 1 mole of benzene

Using Hess's law, the enthalpy of formation for 1 mole of benzene is the sum of the indirect routes.

$$\begin{aligned} \Delta H &= (6 \times \Delta H_1) + (3 \times \Delta H_2) + \Delta H_3 \\ &= (6 \times -394) + (3 \times -286) + (3268) \\ &= (-2364) + (-858) + 3268 \\ &= +46\ \text{kJ mol}^{-1} \end{aligned}$$

Top Tip

−(−3268) becomes + 3268

Standard enthalpy equations

Remember, you must know the definitions of:

- Enthalpy of combustion – enthalpy change when **one mole** of a substance burns completely in oxygen.
- Enthalpy of solution – enthalpy change when **one mole** of a substance dissolves in water.
- Enthalpy of neutralisation – enthalpy change when the acid is neutralised to form **one mole** of water.

Question

1. Methane is formed by the reaction of carbon with hydrogen.

$C(s) + 2H_2(g) \longrightarrow CH_4(g)$

Calculate the enthalpy of formation of methane using enthalpies of combustion from page 9 in your data booklet. Show your working clearly.

The Concept of dynamic equilibrium

Most chemical reactions do not go to completion but exist at equilibrium ($\rightleftharpoons$) i.e.

$A + B \rightleftharpoons C + D$

As reactants (A + B) react to form products (C + D), the products break down to form reactants. The reaction mixture will contain both reactants and products. An equilibrium is reached when the rate of the forward reaction equals the rate of the backward reaction. This equilibrium is called 'dynamic' because the forward and backward reactions take place constantly, even after equilibrium is reached. At equilibrium, the concentrations of reactants and products remain constant, but in most cases the concentrations are not equal. If the concentrations of reactants are less than those of the products, the equilibrium position lies to the product side, and vice versa.

Changes in concentration, temperature and pressure can alter the position of equilibrium by increasing the rate of the forward or backward reactions, resulting in the concentration or reactants and products being altered.

How does concentration affect equilibrium position?

Chlorine dissolves in water: $Cl_2(g) + H_2O(l) \rightleftharpoons 2H^+(aq) + ClO^-(aq) + Cl^-(aq)$

Change in concentration	Effect on products/ reactants	Effect on rate	Effect on equilibrium position
Addition of $Cl_2(g)$ **Addition of reactant**	**Increase** in concentration of **products**.	Rate of forward reaction increases.	Moves to the right.
Addition of HCl(aq) **Addition of product**	**Increase** in concentration of **reactants**	Rate of backwards reaction increases	Moves to the left.
Addition of NaOH **Removal of product**	**Increase** in concentration of **products**.	Rate of forward reaction increases.	Moves to the right.

Alkalis will react with H^+ ions to form water, removing them from the equilibrium mixture.

Acids will remove OH^- ions.

How do temperature and pressure affect equilibrium position?

Consider the following equilibrium: $N_2O_4 \rightleftharpoons 2NO_2$

The forward reaction is an endothermic reaction while the backward reation is exothermic. The equilibrium mixture is a brown gas at 20°C.

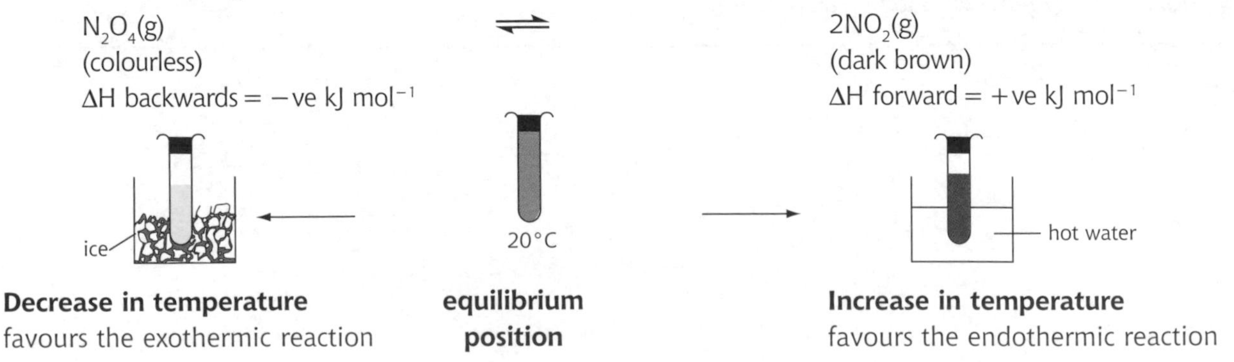

The pressure in a container increases with increasing number of particles in a gas in the container. Pressure will affect any equilibrium which involves gases, if there is a difference between the numbers of moles of gases on the reactant and product sides. The greater the numbers of moles of gas present, the greater the pressure. (If there are the same number of moles of gas, then pressure will have no effect.)

Consider the equilibrium $N_2O_4(g) \rightleftharpoons 2NO_2(g)$

- An increase in pressure favours the direction which leads to fewer molecules, i.e. in this case the backwards reaction.
- A decrease in pressure favours the direction which leads to more molecules, i.e. in this case the forward direction.

Equilibrium and catalyst

A catalyst increases the rate of both the forward and backward reactions, but does not affect the position of equilibrium. The concentration of reactants and products remains the same, but equilibrium is achieved more quickly.

Haber process

Ammonia is produced by the reaction of nitrogen and hydrogen in the presence of an iron catalyst:

$N_2(g) + 3H_2(g) \rightleftharpoons 2NH_3(g)$ $\Delta H = -91 \text{ kJ mol}^{-1}$

four moles (high pressure) two moles (low pressure)

Top Tip

The ΔH value will be for the forward reaction unless otherwise stated in the question.

Conditions required for maximum yield

1. After leaving the reaction chamber, the gaseous mixture is passed through a condenser. The liquid ammonia is constantly removed, reducing the rate of the backward reaction.
2. Unreacted hydrogen and ammonia gases are recycled, increasing the forward reaction.
3. Because the forward reaction is exothermic and has the fewer gas molecules present, the pressure will be low. Low temperatures and high pressure produce the greatest yield.

Chemical plant conditions

The plant operates at higher temperatures than the ideal, because the reaction takes too long at low temperatures. The cost of running a plant at high pressure is high, so the pressure actually used is lower than the ideal pressure.

Questions

1. What effect do catalysts have on equilibrium positions?
2. In the reaction: $2SO_2(g) + O_2(g) \rightleftharpoons 2SO_3(g)$ $\Delta H = -\text{ve}$
   (a) What is the effect on the percentage yield of SO_3 if the pressure is decreased? Explain.
   (b) What would be the effect of decreasing the temperature on the yield of SO_3? Explain.

The pH Scale and calculating the pH of acids and alkalis

The pH scale is a continuous range from below 0 to above 14; pH is a measure of the hydrogen ion concentration in a solution.

Water equilibrium

Water partially dissociates to form hydrogen and hydroxide ions.

$H_2O(l) \rightleftharpoons H^+(aq) + OH^-(aq)$

Water is a poor conductor of electricity because of the small number of ions at equilibrium.

As can be seen from the equation, the concentration of $[H^+]$ to $[OH^-]$ ions is equal. (Concentration of a substance is written using square brackets.) The concentration of each ion is 1×10^{-7} mol l^{-1} at 25°C.

The ionic product of water (Kw) is the multiple of the concentration of H^+ and OH^- ions in a mole of water.

$$Kw = [H^+](aq) \times [OH^-](aq)$$
$$= 1 \times 10^{-7} \text{ mol l}^{-1} \times 1 \times 10^{-7} \text{ mol l}^{-1}$$
$$= 1 \times 10^{-14} \text{ mol}^2 \text{ l}^{-2}$$

Equilibrium in acids and alkalis

The ionic product of water always remains at 10^{-14} mol^2 l^{-2} but the ratio of H^+:OH^- can change.

An **acid** is a hydrogen ion donor: it releases hydrogen ions when dissolved in water. This results in an increase in H^+ ions and a decrease in OH^- ions.

If an **alkali** is dissolved in water, the concentration of OH^- ions increases, resulting in a decrease in the concentration of H^+ ions.

Calculating the pH of acids and alkalis

pH is expressed as the negative log of the H^+ concentration: $pH = -\log_{10} [H^+(aq)]$

1. What is the pH of a 0.1 mol l^{-1} solution of HCl?

Step 1 Express the concentration of H^+ ions as $1 \times 10^X = 1 \times 10^{-1}$

Step 2 Calculate pH $= -\log_{10} [H^+]$

$= -\log_{10} [1 \times 10^{-1}]$

$= 1$

2. What is the pH of a 0.1 mol l^{-1} solution of NaOH?

Step 1 Express the concentration of OH^- ions as 10^X $= 1 \times 10^{-1}$

Step 2 Calculate the number of H^+ ions present $[H^+] \times [OH^-] = 1 \times 10^{-14}$ mol^2 l^{-2}

$[H^+] = \frac{1 \times 10^{-14}}{[OH^-]}$ (because $[H^+][OH^-] = 10^{-14}$)

$[H^+] = \frac{1 \times 10^{-14}}{1 \times 10^{-1}} = 10^{-13}$

$pH = -\log[1 \times 10^{-13}] = 13$

Top Tip

$[H^+] \times [OH^-] = 1 \times 10^{-14}$ mol^2 l^{-2}

Convert to 10^X as follows:

$0.01 = 1 \times 10^{-2}$

$0.001 = 1 \times 10^{-3}$

$0.0001 = 1 \times 10^{-4}$

Summary of pH values

A summary of pH values for different concentrations of acids and alkalis is shown below.

Concentration acid (mol l^{-1})	Concentration of H^+ ions (1×10^{-x} mol l^{-1})	pH	Concentration of OH^- ions (1×10^{-x} mol l^{-1})	Concentration of alkali (mol l^{-1})
1	1×10^{0}	0	1×10^{-14}	0.00000000000001
0.1	1×10^{-1}	1	1×10^{-13}	0.0000000000001
0.01	1×10^{-2}	2	1×10^{-12}	0.000000000001
0.001	1×10^{-3}	3	1×10^{-11}	0.00000000001
0.0001	1×10^{-4}	4	1×10^{-10}	0.0000000001
0.00001	1×10^{-5}	5	1×10^{-9}	0.000000001
0.000001	1×10^{-6}	6	1×10^{-8}	0.00000001
0.0000001	1×10^{-7}	7	1×10^{-7}	0.0000001
0.00000001	1×10^{-8}	8	1×10^{-6}	0.000001
0.000000001	1×10^{-9}	9	1×10^{-5}	0.00001
0.0000000001	1×10^{-10}	10	1×10^{-4}	0.0001
0.00000000001	1×10^{-11}	11	1×10^{-3}	0.001
0.000000000001	1×10^{-12}	12	1×10^{-2}	0.01
0.0000000000001	1×10^{-13}	13	1×10^{-1}	0.1
0.00000000000001	1×10^{-14}	14	1×10^{0}	1

Example

State the concentration of hydroxide ions in a solution with a pH of 4

Step 1 Convert pH into 1×10^{x} pH = 4 $= 1 \times 10^{-4}$

Step 2 Calculate the number of OH^- ions $[H^+] \times [OH^-] = 1 \times 10^{-14}\ mol^2\ l^{-2}$

$[OH^-] = 1 \times 10^{-14} / [H^+]$

$= 1 \times 10^{-14} / 10^{-4}$

$= 1 \times 10^{-10}\ mol\ l^{-1}$

Questions

1. Why is water a poor conductor of electricity?
2. Calculate the concentration of H^+ ions present in a solution with a pH of 3.
3. Calculate the pH of a 0.001 mol l^{-1} solution of sodium hydroxide.
4. What is the concentration of OH^- ions in a solution with a pH of 5?
5. Using Avogadro's constant, calculate the number of H^+ ions in 100 cm^3 of a 0.1 mol l^{-1} solution of HCl.

Strong and weak acids and bases

Strong acids

When a strong acid is dissolved in water, it is fully **dissociated** (i.e. ionised). HCl is a strong acid:

$HCl(aq) \longrightarrow H^+(aq) + Cl^-(aq)$

This full dissociation is due to the polar covalent nature of the acid molecule. Full dissociation of the acid molecule is shown by a single arrow. Strong acids include $HCl(aq)$, $HNO_3(aq)$ and $H_2SO_4(aq)$.

Weak acids

A weak acid only partially dissociates, resulting in an equilibrium existing between the acid molecule and its ions. This is shown by the $\rightleftharpoons$ arrow.

Alkanoic acids such as ethanoic acid are examples of weak acids.

$$H_3C{-}C(=O){-}O{-}H \rightleftharpoons H_3C{-}C(=O){-}O^-(aq) + H^+(aq)$$

Carbonic acid (CO_2 dissolved in H_2O) and sulphurous acid (SO_2 dissolved in H_2O) are further examples of weak acids. Their weak nature can be illustrated by the following equilibrium reactions:

$H_2CO_3(aq) \rightleftharpoons 2H^+(aq) + CO_3^{2-}(aq)$

$H_2SO_3(aq) \rightleftharpoons 2H^+(aq) + SO_3^{2-}(aq)$

Comparing weak and strong acids

The following results were obtained when solutions of a strong and a weak acid of the same concentration were compared.

Test	**Strong acid (0.1 mol l^{-1} HCl)**	**Weak acid (0.1 mol l^{-1} CH_3COOH)**
pH	1	2.8
Conductivity	High	Low
Reaction with Mg	Fast reaction: $H_2(g)$ released	Slow reaction: $H_2(g)$ released
Titration with NaOH(aq)	The HCl reacts with 25 cm^3 of 0.1 mol l^{-1} NaOH	The CH_3COOH reacts with 25 cm^3 of 0.1 mol l^{-1} NaOH

pH – pH is a measure of H^+ ions. Therefore a strong acid will have a low pH because it gives a greater concentration H^+ ions than weak acids.

Conductivity – The conductivity of a solution increases with the number of ions present.

Strong acids will always have higher conductivities due to the greater numbers of ions present.

Reactions – Strong acids fully dissociate and therefore have excess H^+ ions which react. Weak acids do not fully dissociate but as the H^+ ions react, the equilibrium position changes and more H^+ ions are released to react. In both cases, the acids release all their stock of H^+ ions so they both give the same volume or mass of product.

Strong bases

A base is a substance which accepts hydrogen ions and neutralises an acid. Bases which dissolve in water are called **alkalis**. When a strong base dissolves in water, all of its OH^- ions are released into solution. This is shown by a single arrow.

$NaOH(s) \longrightarrow Na^+(aq) + OH^-(aq)$

Strong bases include NaOH, LiOH and KOH.

Weak bases

A weak base only partially dissociates, resulting in an equilibrium existing between the molecule and its ions. This occurs when ammonia dissolves in water to form ammonium ions and hydroxide ions. It is shown as follows:

$NH_3(g) + H_2O(l) \rightleftharpoons NH_4 + (aq) + OH^-(aq)$

Comparing weak and strong alkalis

Test	**0.1 mol l^{-1} NaOH(aq)**	**0.1 mol l^{-1} NH_3(aq)**
pH	14	10
Conductivity	High	Low
Titration with NaOH(aq)	The NaOH reacts with 25 cm^3 of 0.1 mol l^{-1} HCl	The NH_3 reacts with 25 cm^3 of 0.1 mol l^{-1} HCl

pH – A strong alkali has a high pH because it contains more OH^- than a weak alkali, due to complete dissociation.

Conductivity – The conductivity of a solution increases with the number of ions present.
A strong alkali will have a higher pH than a weak alkali due to the greater number of ions present.

Reactions – Strong alkalis or bases fully dissociate and therefore have excess OH^- ions which react. Weak alkalis or bases do not fully dissociate but as the OH^- ions react, the equilibrium position changes and more OH^- ions are released to react.

In both cases, the alkalis release all their stock of OH^- ions so they both give the same volume or mass of product.

Questions

1. Nitric acid is an example of a strong acid. Using equations explain why HNO_3 is a strong acid.
2. Describe three tests that would show the difference between NH_4OH(aq) and KOH(aq). Explain the results you would obtain.
3. Using equations, show why ethanoic acid is an example of a weak acid.

The pH of salt solutions

A soluble salt of a **strong acid** and **strong base** dissolves in water to produce a **neutral** solution.

A soluble salt of a **weak acid** and **strong base** dissolves in water to produce an **alkaline** solution.

A soluble salt of a **strong acid** and a **weak base** dissolves in water to produce an **acidic** solution.

Why is a given salt acidic?

The salt ammonium chloride dissolves in water, to form an acidic solution.

Using appropriate equilibria explain why this is so.

Step 1 – When a salt dissolves in water, it fully dissociates to form its ions. Water partially dissociates to form hydrogen and hydroxide ions. A solution of NH_4Cl contains H_2O molecules, H^+ ions, OH^- ions, NH_4^+ ions and Cl^- ions.

The first step therefore is to write the equations explaining the above.

Full dissociation of salt: $NH_4Cl \longrightarrow NH_4^+ + Cl^-$

Partial dissociation of water: $H_2O \rightleftharpoons OH^- + H^+$

Step 2 – Select the ion which originates from the weak source.

In this case, the **ammonium ion** originates from the weak alkali ammonium hydroxide. The ammonium ions will react with the hydroxide ions provided by the dissociation of water.

Bring all the equations together.

$$
\begin{array}{lll}
NH_4Cl(aq) & \longrightarrow & NH_4^+(aq) \ + Cl^-(aq) \\
H_2O(l) & \rightleftharpoons & OH^-(aq) \ + H^+(aq) \\
 & & \quad \upharpoonleft\downharpoonright \\
 & & NH_3(aq) \\
 & & + \\
 & & H_2O(l)
\end{array}
$$

When writing the water equilibrium, always place OH^- ions below the ions from the alkaline source. Place the H^+ ions below the ions from the acid source.

Step 3 – Learn the following explanation:

The $OH^-(aq)$ ions react with the $NH_4^+(aq)$ to form ammonia, $NH_3(aq)$.

Hydroxide ions are removed from the water equilibrium and must be replaced.

Water molecules break down to replace the lost hydroxide ions and produce hydrogen ions at the same time, resulting in an excess of hydrogen ions and the formation of an acidic solution.

Summary of steps

Step 1 – Write the equations for

a) partial dissociation of water

b) full dissociation of the salt.

Step 2 – Draw an arrow to show that the ion from the weak alkali reacts with the OH^- ion from the partial dissociation of water. This reaction is reversible, so use the equilibrium arrow.

Step 3 – Explain that the removal of OH^- ions drives the water equilibrium to the right, to replace these lost OH^- ions. This produces H^+ ions at the same time, so forming an acidic solution.

Why is a given salt alkaline?

Sodium ethanoate dissolves in water to form an alkaline solution.

Step 1 – Write equations to show the full dissociation of salt and partial dissociation of water. Draw all the equations together.

Full dissociation of salt $CH_3COONa(aq) \longrightarrow CH_3COO^-(aq) + Na^+(aq)$

+

Partial dissociation of water $H_2O(l) \rightleftharpoons H^+(aq) + OH^-(aq)$

$\updownarrows$

$CH_3COOH(aq)$

Top Tip

Draw all the equations together.

Step 2 – Draw a line to show that the ion from the weak acid reacts with the hydrogen ion in the water equilibrium. This reaction is reversible.

Step 3 – H^+ ions react with the ethanoate ions to form ethanoic acid.
Water molecules break down to replace the lost hydrogen ions and produce hydroxide ions at the same time. Therefore an excess of hydroxide ions is formed and the solution is alkaline.

Soaps

Soaps are formed by the hydrolysis of fats and oils. Fats and oils are boiled in sodium hydroxide to produce propan-1,2,3-triol (glycerol) and sodium salts of fatty acids.

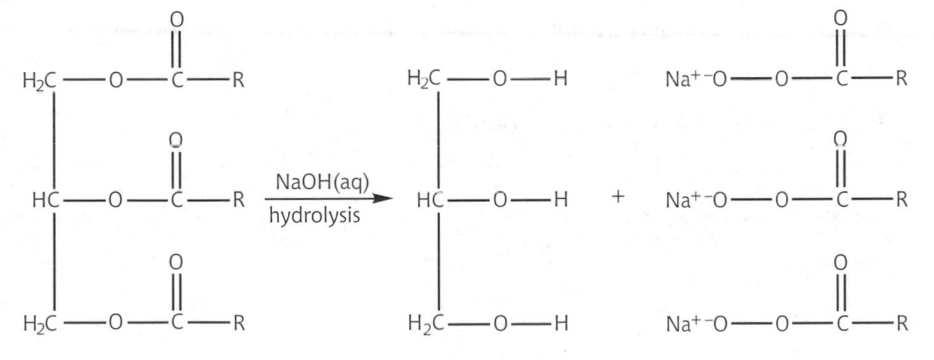

fat or oil molecule propane-1,2,3-triol sodium salts of fatty acids

The fatty acids are carboxylic acids ranging from C_4 to C_{24}. Carboxylic acids are weak acids, so soaps are salts formed from a weak acid and a strong base. Soaps dissolve in water to form **alkaline solutions**.

Questions

1. Classify the following salts as acidic, alkaline or neutral:
   (*a*) ammonium chloride
   (*b*) sodium ethanoate
   (*c*) sodium chloride
2. Using equations, explain why ammonium nitrate is an acidic salt.

Oxidising and reducing agents

During a **redox** reaction, one species will lose electrons (oxidation), while another species will gain electrons (reduction).

Displacement reactions are examples of redox reactions.

Writing redox reactions

Example

Magnesium reacts with a solution of silver(I) ions to form silver and a solution of magnesium(II) ions.

Step 1 – List reactants and products.

Reactants: Mg and Ag^+ Products: Mg^{2+} and Ag

Step 2 – Write the reduction reactions found in the data booklet (page 11).

$$Mg^{2+}(aq) + 2e^- \longrightarrow Mg(s)$$

$$Ag^+(aq) + e^- \longrightarrow Ag(s)$$

Step 3 – Rearrange equations to show an oxidation reaction and a reduction reaction.

Remember to look at your reactants and products in step 1 to see which equation is reversed.

Oxidation $Mg(s) \longrightarrow Mg^{2+}(aq) + 2e^-$

(Reaction is reversed because Mg is a reactant and not a product.)

Reduction $Ag^+(aq) + e^- \longrightarrow Ag(s)$

Step 4 – Multiply each equation so that electrons are equal in number.

The silver equation is multiplied by two.

Oxidation $Mg(s) \longrightarrow Mg^{2+}(aq) + 2e^-$

Reduction $2Ag^+(aq) + 2e^- \longrightarrow 2Ag(s)$

Step 5 – Add the reduction and oxidation reactions together, eliminating the electrons.

Redox $Mg(s) + 2Ag^+(aq) \longrightarrow Mg^{2+}(aq) + 2Ag(s)$

Top Tip

Arrange equations so that reactants are under reactants and products are under products.

Reducing and oxidising agents

- An **oxidising agent** is a substance which accept electrons from another substance. The oxidising agent is reduced and the other species is oxidised.
- A **reducing agent** is the opposite of an oxidising agent, i.e. it donates electrons to another species. The reducing agent is oxidised and the other species is reduced.
  In the above example, Mg is a reducing agent and Ag^+ is an oxidising agent.

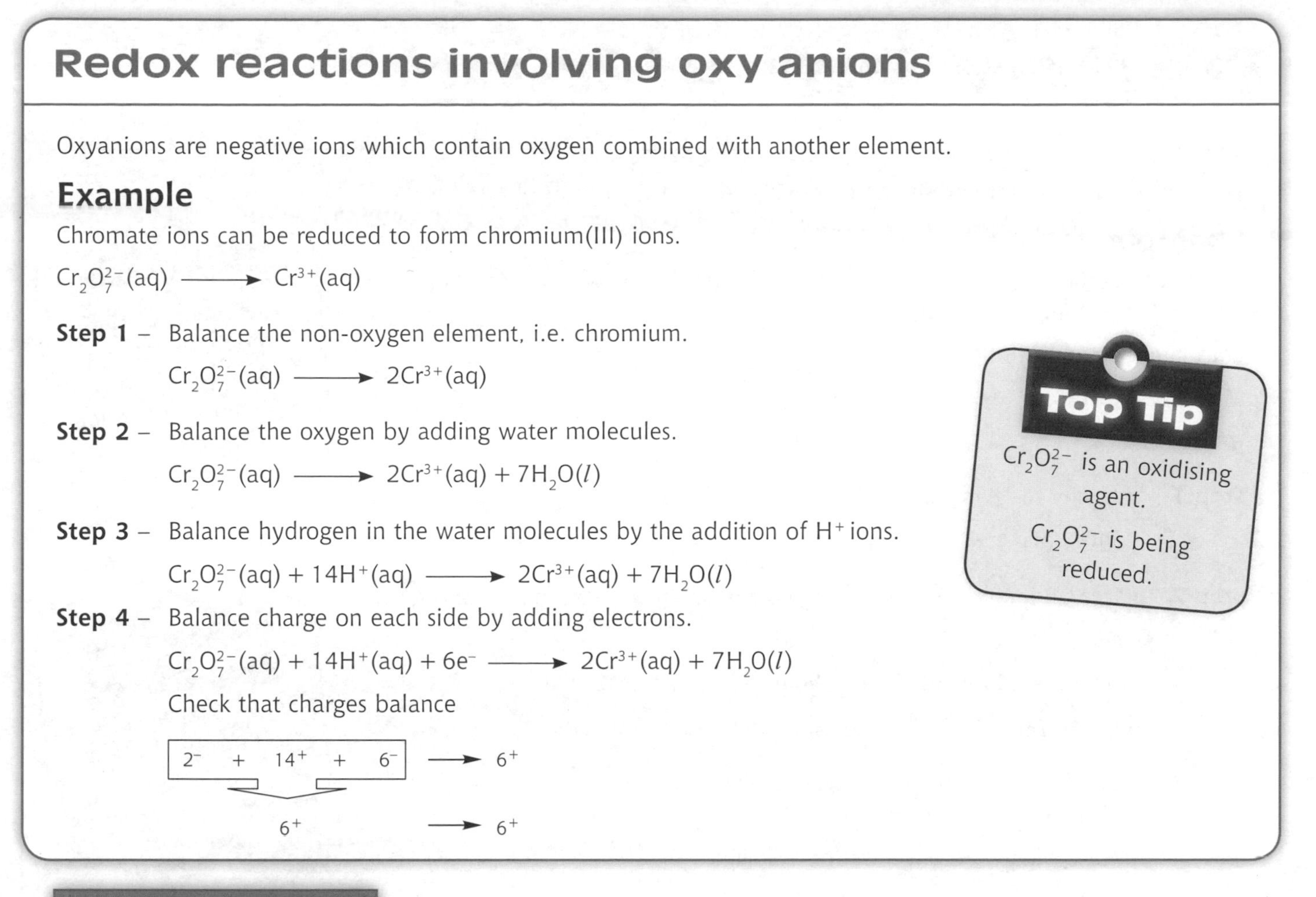

Redox reactions involving oxy anions

Oxyanions are negative ions which contain oxygen combined with another element.

Example

Chromate ions can be reduced to form chromium(III) ions.

$Cr_2O_7^{2-}(aq) \longrightarrow Cr^{3+}(aq)$

Step 1 – Balance the non-oxygen element, i.e. chromium.

$Cr_2O_7^{2-}(aq) \longrightarrow 2Cr^{3+}(aq)$

Step 2 – Balance the oxygen by adding water molecules.

$Cr_2O_7^{2-}(aq) \longrightarrow 2Cr^{3+}(aq) + 7H_2O(l)$

Step 3 – Balance hydrogen in the water molecules by the addition of H^+ ions.

$Cr_2O_7^{2-}(aq) + 14H^+(aq) \longrightarrow 2Cr^{3+}(aq) + 7H_2O(l)$

Step 4 – Balance charge on each side by adding electrons.

$Cr_2O_7^{2-}(aq) + 14H^+(aq) + 6e^- \longrightarrow 2Cr^{3+}(aq) + 7H_2O(l)$

Check that charges balance

$2^- + 14^+ + 6^- \longrightarrow 6^+$

$6^+ \longrightarrow 6^+$

Top Tip

$Cr_2O_7^{2-}$ is an oxidising agent.

$Cr_2O_7^{2-}$ is being reduced.

Questions

1. Aluminium displaces copper from a solution of copper(II) sulphate. Write the redox reaction for this displacement reaction.
2. Identify the oxidising and reducing agents in the above reaction.
3. Write the ion electron equation for the reduction of nitrate (NO_3^-) ions to nitrogen monoxide (NO).
4. Write the ion electron equation for the reduction of $FeO_4^{2-}(aq)$ to $Fe^{3+}(aq)$.

Redox titrations

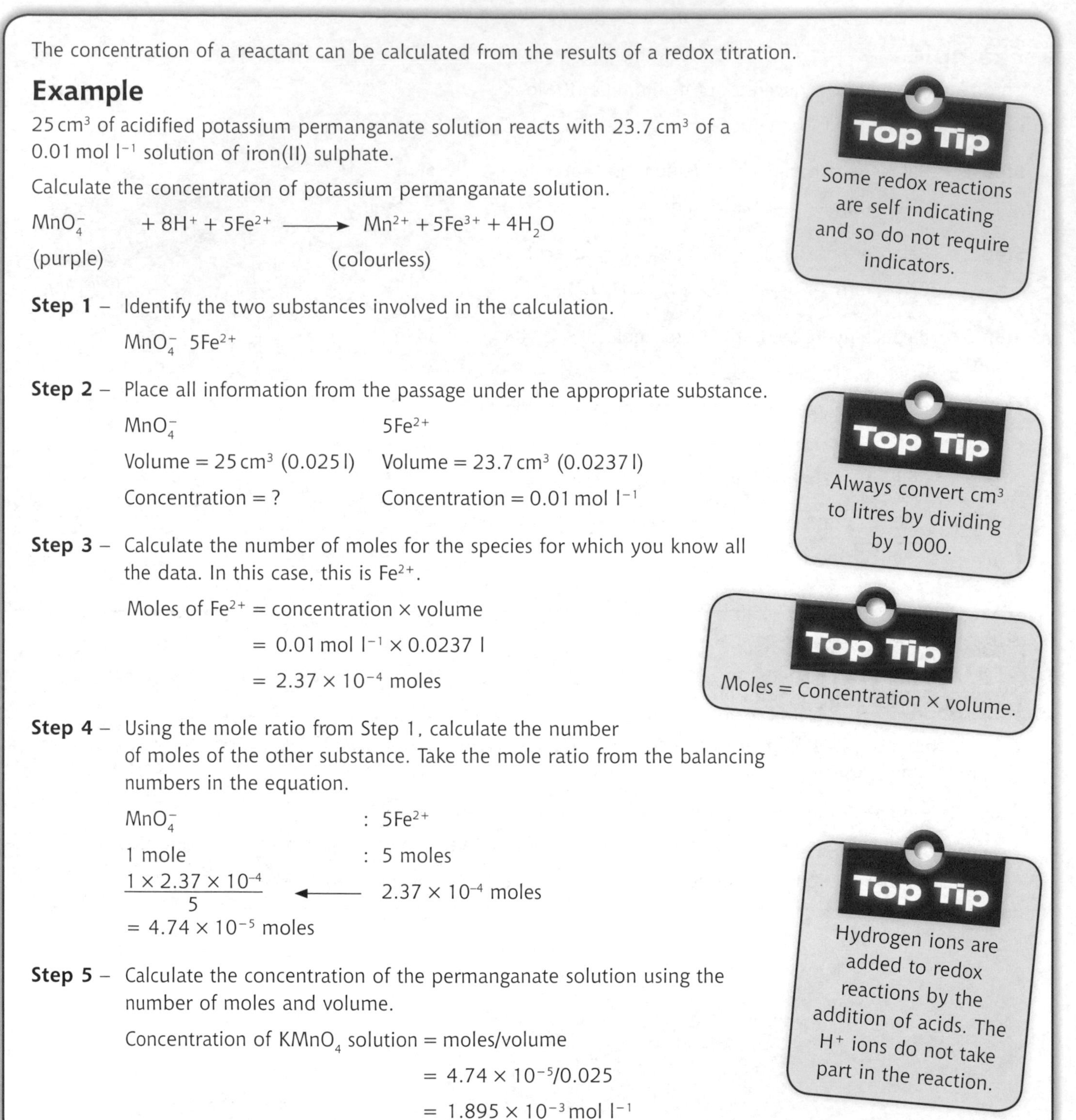

The concentration of a reactant can be calculated from the results of a redox titration.

Example

25 cm³ of acidified potassium permanganate solution reacts with 23.7 cm³ of a 0.01 mol l^{-1} solution of iron(II) sulphate.

Calculate the concentration of potassium permanganate solution.

MnO_4^- (purple) $+ 8H^+ + 5Fe^{2+} \longrightarrow Mn^{2+}$ (colourless) $+ 5Fe^{3+} + 4H_2O$

Step 1 – Identify the two substances involved in the calculation.

MnO_4^- $5Fe^{2+}$

Step 2 – Place all information from the passage under the appropriate substance.

MnO_4^-	$5Fe^{2+}$
Volume = 25 cm³ (0.025 l)	Volume = 23.7 cm³ (0.0237 l)
Concentration = ?	Concentration = 0.01 mol l^{-1}

Step 3 – Calculate the number of moles for the species for which you know all the data. In this case, this is Fe^{2+}.

Moles of Fe^{2+} = concentration × volume

$= 0.01 \text{ mol } l^{-1} \times 0.0237 \text{ l}$

$= 2.37 \times 10^{-4}$ moles

Step 4 – Using the mole ratio from Step 1, calculate the number of moles of the other substance. Take the mole ratio from the balancing numbers in the equation.

MnO_4^- : $5Fe^{2+}$

1 mole : 5 moles

$\frac{1 \times 2.37 \times 10^{-4}}{5}$ ⟵ 2.37×10^{-4} moles

$= 4.74 \times 10^{-5}$ moles

Step 5 – Calculate the concentration of the permanganate solution using the number of moles and volume.

Concentration of $KMnO_4$ solution = moles/volume

$= 4.74 \times 10^{-5}/0.025$

$= 1.895 \times 10^{-3} \text{ mol } l^{-1}$

Top Tip

Some redox reactions are self indicating and so do not require indicators.

Top Tip

Always convert cm³ to litres by dividing by 1000.

Top Tip

Moles = Concentration × volume.

Top Tip

Hydrogen ions are added to redox reactions by the addition of acids. The H^+ ions do not take part in the reaction.

PPA Three: A Redox reaction

Introduction

Vitamin C (ascorbic acid) can be obtained through a healthy diet that includes fruits and vegetables.

Some people supplement their diets by taking Vitamin C tablets.

The aim of this experiment is to determine the mass of vitamin C in a tablet by redox titration, using a solution of iodine of accurately known concentration and starch solution as an indicator. The redox equation is

$C_6H_8O_6(aq) + I_2(aq) \longrightarrow C_6H_6O_6(aq) + 2H^+(aq) + 2I^-(aq)$
(ascorbic acid)

Procedure

- A vitamin C tablet was dissolved in a small volume of deionised water. This solution was then added to a 250 cm^3 standard flask. The beaker was then washed with water. The washings were added to the flask which was then made up to 250 cm^3 with water.
- 25 cm^3 of the vitamin C solution was pipetted into a conical flask, along with a few drops of starch solution.
- The solution was titrated against a known concentration of iodine until a blue/black solution was formed. This titration was repeated until the results were concordant (within 0.1 cm^3 of each other).

Top Tip

Starch is used as an indicator in redox reactions that involve iodine. Starch is blue/black in the presence of $I_2(aq)$ and colourless in the presence of $2I^-(aq)$.

Example results

The concentration of iodine solution was 0.034 mol l^{-1}.

Gram formula mass of vitamin C is 176 g.

Titre One = 18.9 cm^3

Titre Two = 16.5 cm^3

Titre Three = 16.4 cm^3

Average Titre = 16.45 cm^3 = 0.01645 l

Top Tip

Average titre should not include rough titre.

Step 1 – Moles of iodine = C × V
= 0.034 × 0.01645
= 5.593×10^{-4} moles

Step 2 – Mole ratio I_2 : $C_6H_8O_6$
1 mole : 1 mole
5.593×10^{-4} : 5.593×10^{-4}

Step 3 – Scale up moles of vitamin C: = 5.593×10^{-4} moles in a 25 cm^3 sample
= 5.593×10^{-3} moles in 250 cm^3 moles

Step 4 – Determine mass of vitamin C: n × gfm
= $5.593 \times 10^{-3} \times 176$
= 0.984 g

Top Tip

Always remember to scale up if titration is a proportion of the total volume.

Questions

1. Fe^{2+} ions are oxidised to Fe^{3+} by chromate ions: $6Fe^{2+} + Cr_2O_7^{2-} + 14H^+ \longrightarrow 6Fe^{3+} + 2Cr^{3+} + 7H_2O$

   Calculate the concentration of a Fe^{3+} solution if 25 cm^3 reacted with 20 cm^3 of 1 mol l^{-1} $Cr_2O_7^{2-}$ solution.

What Is electrolysis?

When a current of electricity is passed through an ionic solution, the compound is broken down to produce its elements. The electrolysis of $CuCl_2(aq)$ produces copper and chlorine.

Negative electrode

$Cu^{2+}(aq) + 2e \longrightarrow Cu(s)$

Positive electrode

$2Cl^-(aq) \longrightarrow Cl_2(g) + 2e^-$

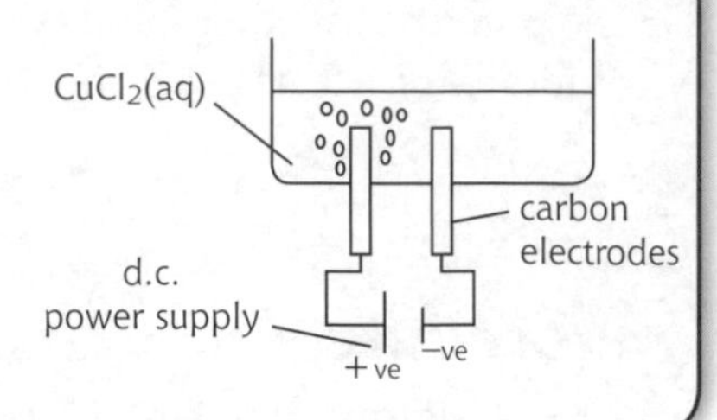

A reduction reaction takes place at the negative electrode while an oxidation reaction takes place at the positive electrode.

Quantitative electrolysis

Faraday's number

Faraday's law states that the number of moles of a substance produced at an electrode during electrolysis is proportional to the number of moles of electrons transferred to the electrode.

The amount of electrical charge carried by one mole of electrons is 96500C. This is called Faraday's symbol (F).

Therefore:

$Na^+(aq) + e^- \longrightarrow Na(s)$

One mole of electrons produces one mole of sodium **1** × 96 500C

$Mg^{2+}(aq) + 2e^- \longrightarrow Mg(s)$

Two moles of electrons produce one mole of magnesium **2** × 96 500C

$Al^{3+}(aq) + 3e^- \longrightarrow Al(s)$

Three moles of electrons produce one mole of aluminium **3** × 96 500C

Calculating the charge

The number of coulombs of charge going through a solution or melt can be calculated using the following formula:

Q (coulombs) = I (current in amps) × t (time in seconds)

Calculating the mass or volume produced

Example

Calculate the mass of aluminium produced if a current of 5 amps is passed through a melt of Al_2O_3 for 4 minutes and 20 seconds.

Step 1 – Calculate the number of coulombs of charge

$Q = I \times t$

$= 5 \times 260$

$= 1300$ C

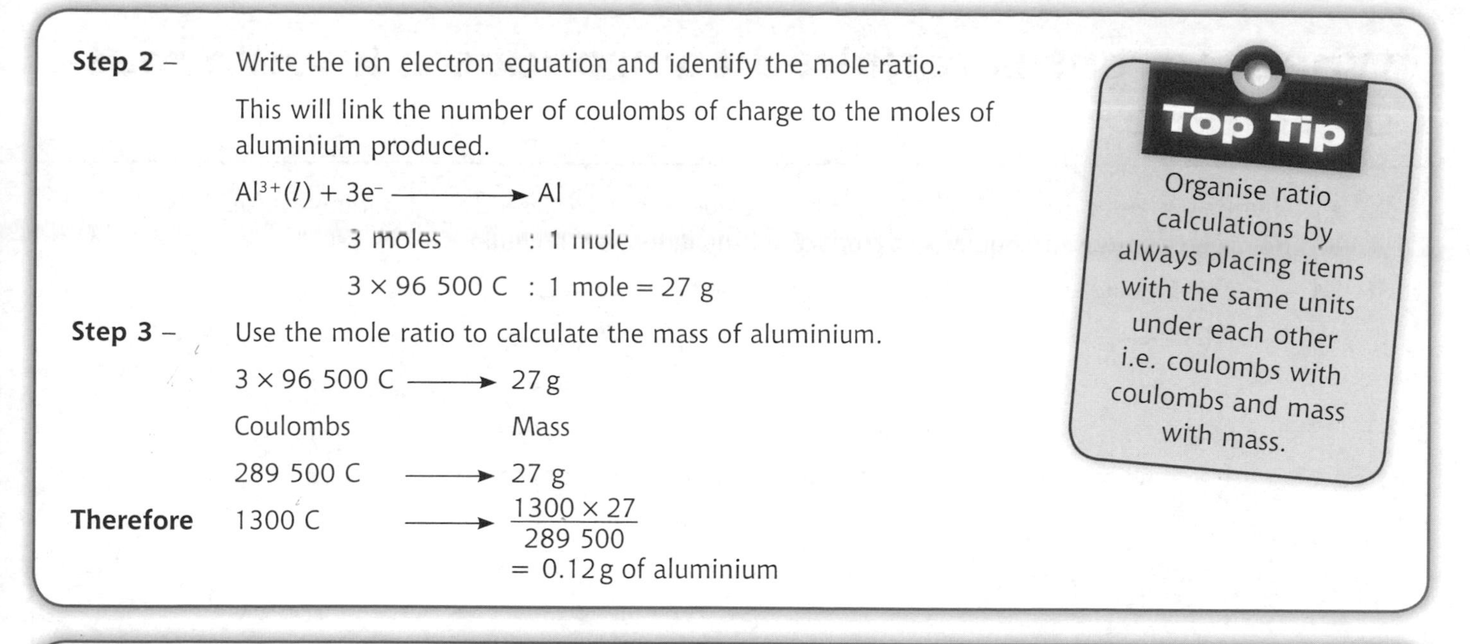

Step 2 – Write the ion electron equation and identify the mole ratio.

This will link the number of coulombs of charge to the moles of aluminium produced.

$Al^{3+}(l) + 3e^- \longrightarrow Al$

3 moles : 1 mole

3 × 96 500 C : 1 mole = 27 g

Step 3 – Use the mole ratio to calculate the mass of aluminium.

3 × 96 500 C ⟶ 27 g

Coulombs Mass

289 500 C ⟶ 27 g

Therefore 1300 C ⟶ $\frac{1300 \times 27}{289\ 500}$

= 0.12 g of aluminium

Top Tip

Organise ratio calculations by always placing items with the same units under each other i.e. coulombs with coulombs and mass with mass.

What time is required to produce a known mass or volume?

A current of 5 amps was passed through a solution of $CuCl_2$ for a period of time. Calculate the time taken to produce 221 cm^3 of chlorine gas, given that the molar volume of chlorine is 22.1 litres mol^{-1}.

Step 1 – Write the ion electron equation occurring and determine the mole ratio.

$2Cl^-(aq) \longrightarrow Cl_2(g) + 2e$

1 mole : 2 moles

22.1 l : 2 × 96 500 C (= 193 000 C)

Step 2 – Use the mole ratio to determine the number of coulombs of charge.

22.1 l ⟶ 193 000 C

22 100 cm^3 ⟶ 19 300 C

Therefore 221 cm^3 ⟶ $\frac{221 \times 19\ 300}{22\ 100}$

= 1930 C

Step 3 – Rearrange Q = I × t, to calculate the time taken to produce 221 cm^3 of chlorine gas.

t = Q ÷ I

= 1930 ÷ 5

= 386 s

= 6 minutes and 26 seconds

What current is required to produce a known mass or volume

2.7 g of aluminium was produced when a melt of its oxide was electrolysed for 25 minutes 40 secondss (1540s). Calculate the current required to produce this quantity of aluminium.

Step 1 – Write the ion electron equation and determine the mole ratio.

$Al^{3+}(l)$ + $3e^-$ ⟶ $Al(l)$

3 moles : 1 mole

3 × 96 500 C = 289 500 C : 1 mole = 27g

Step 2 – Use the mole ratio to determine the number of coulombs of charge.

27 g → 289 500 C

Therefore 2.7 g → $\frac{2.7 \times 289\,500}{27}$

= 28 950 C

Step 3 – Rearrange Q = I × t, to calculate the current taken to produce

I = Q ÷ t

= 28 950 C ÷ 1540s

= 18.8 A

Questions

1. What mass of copper will be produced when a solution of copper(II) sulphate is electrolysed for 30 minutes and 40 seconds with a constant current of 5 amps?
2. A current of 3 amps was passed through a melt of aluminium oxide for 15 minutes and 5 seconds. Calculate the mass of aluminium produced.
3. 12 A was passed through a solution of NaCl. Calculate the time required to produce 3.8g of sodium.
4. Calculate the time required to produce 12g of magnesium from a melt of its oxide. The current used was 5 amps.
5. What current was passed through a solution of copper(II) chloride to produce 2.12g of copper? The time taken for the electrolysis was 20 min and 50 s.
6. What current is required to produce 100 cm^3 of hydrogen gas from the electrolysis of aqueous hydrochloric acid, if the time taken is 50 minutes and the molar volume of hydrogen gas is 22.1 $l\ mol^{-1}$?
7. What volume of chlorine gas is produced when a current of 2.5 A is passed through a solution of copper chloride for 10 minutes and 10 seconds. Assume the molar volume for chlorine is 24 $l\ mol^{-1}$.

PPA – Quantitative electrolysis

Introduction

During electrolysis, hydrogen ions in sulphuric acid are reduced to hydrogen gas.

$2H^+(aq) + 2e \longrightarrow H_2(g)$

The ion electron equation above shows that two moles of electrons are required to liberate one mole of hydrogen. Because 96 500 C of charge is associated with one mole of electrons, then 2 × 96 500 C will be needed to liberate one mole of hydrogen.

The aim of this is to determine the quantity of electricity required to produce one mole of hydrogen by electrolysing dilute sulphuric acid.

Procedure

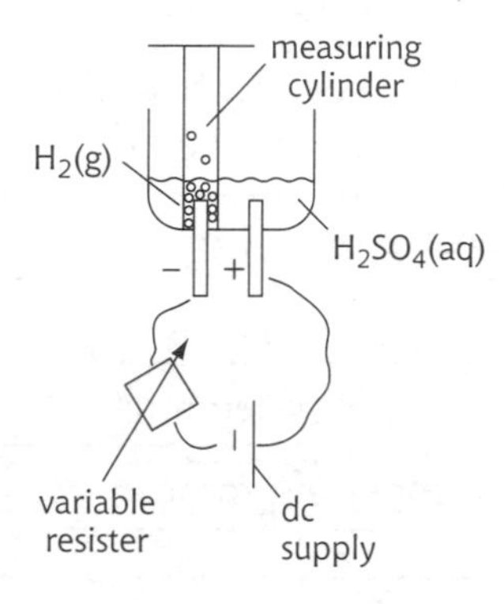

- The apparatus opposite was set up.
- Dilute sulphuric acid was added to the cell to just above the height of the electrodes.
- A measuring cylinder was then filled with dilute sulphuric acid, inverted and placed underneath the surface of the acid in the cell.
- The lab pack was switched on and the variable resistor altered until a current of 0.5 amps was obtained. The current was passed through the solution for a few minutes and then switched off.
- The current was switched on and the time taken for 50 cm³ of gas to be collected was recorded. The variable resistor was adjusted throughout to ensure a constant current of 0.5 amps.

Example results

The time taken to collect 50 cm³ of hydrogen was 13 minutes and 26 seconds (806 s).

Quantity of electricity used = I × t

= 0.5 A × 806 s

= 403 C

The molar volume of hydrogen at room temperature is 24 litres mol⁻¹ (24 000 cm³).

Since 50 cm³ of H_2 = 403 C

24000 cm³ $= \frac{24\,000\ cm^3 \times 403\ C}{50\ cm^3}$

= 193 440 C

Conclusion

The experimental value was close to the theoretical value of 193 000 C

Evaluation

Possible sources of error:

1. The current could have fluctuated. This was minimised by the use of a variable resistor.
2. Some of the hydrogen could have dissolved into the sulphuric acid during electrolysis. The solution was electrolysed for a while before the timer was started, to allow the porous carbon electrode to become saturated in hydrogen gas so that the volume of $H_2(g)$ recorded was not underestimated.

Question

1. Why was a variable resistor used in the PPA 'electrolysing sulphuric acid'?

Types of radiation

What are isotopes?

Atoms are made up of three subatomic particles: protons, neutrons and electrons.

Protons and neutrons are the subatomic particles which make up the nucleus of an atom.

The number of protons and neutrons in an atom can be determined using the atomic and mass numbers.

- Number of protons = atomic number
- Number of neutrons = mass number – atomic number

Isotopes are atoms of the same element with differing mass numbers and therefore different numbers of neutrons. Hydrogen has three isotopes.

Name of isotope	**Protium**	**Deuterium**	**Tritium**
$^{\text{mass number}}_{\text{atomic number}}$Symbol	$^{1}_{1}H$	$^{2}_{1}H$	$^{3}_{1}H$
Number of neutrons	0	1	2
Number of protons	1	1	1

What are radioisotopes?

The stability of a nucleus in an atom depends on the proton:neutron ratio.

Stable nuclei (lighter elements) contain roughly equal numbers of protons and neutrons.

As the nuclei of the elements increase in size, the ratio of neutrons to protons increases.

Nuclei of the heavier elements are therefore unstable. Most of the isotopes of elements beyond element 83 are unstable.

Radioactivity is the result of unstable nuclei rearranging to form stable nuclei.

Radioisotopes are unstable nuclei which emit radiation and energy to form stable nuclei.

Background radiation

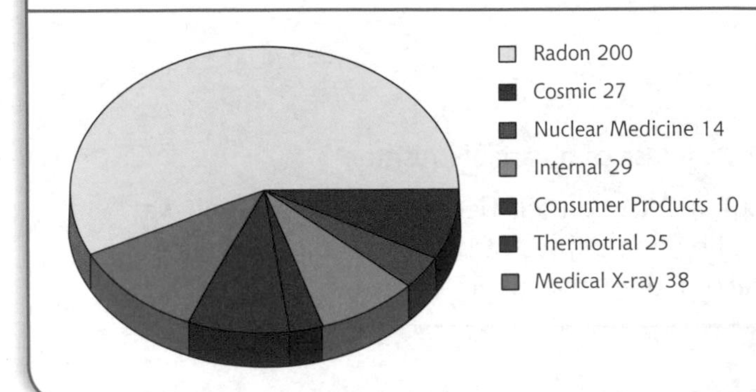

The world around us contains many radioactive sources and therefore we are exposed to radiation all the time.

This is called background radiation.

These sources range from radioactive elements which make up some forms of rocks (radon), building materials, cosmic rays, medical applications and disposal of nuclear waste, to items in the home, such as smoke detectors. The proportions of background radiation are shown on this pie chart.

Types of radiation

There are three different types of radiation: **alpha**, **beta** and **gamma**. Each type of radiation has different properties from the others.

Radiation	Symbol	Composition	Charge/deflection
Alpha	α	Helium atoms: ^4_2He	Have positive charge, strongly deflected by negative charge
Beta	β	Produced when a neutron breaks down in the nucleus to produce a proton and electron (beta particle). $^1_0n \longrightarrow {^1_1p} + {^0_1e}$ The proton stays in the nucleus. The electron is emitted.	Have negative charge, slightly deflected by positive charge
Gamma	γ	Consists of a high-energy electromagnetic wave.	Have no charge. No deflection

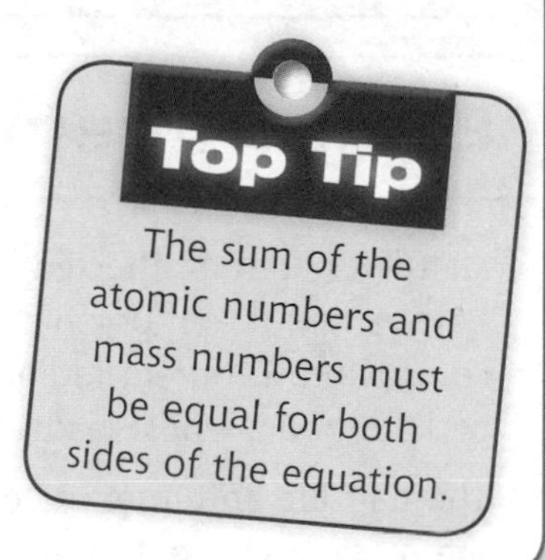

Alpha, beta and gamma radiation have different penetrating properties. As shown above, whilst alpha particles can be stopped by a few centimetres of air, it requires a thin sheet of aluminium to stop beta particles. The high-energy waves of gamma radiation require thick layers of lead or concrete.

Nuclear equations

When a radioactive isotope (radioisotope) disintegrates, it emits a particular type of radiation. The change that occurs inside the nuclei depends on the radiation emitted.

Alpha-emitting isotopes

$$^{222}_{86}Rn \longrightarrow {^{218}_{84}Po} + {^4_2\alpha}$$

Mass number decreases by four, atomic number decreases by two.

Beta-emitting isotopes

$$^{14}_6C \longrightarrow {^{14}_7N} + {^0_{-1}e}$$

Atomic number increases by one, no change to mass number.

Gamma-emitting isotopes

$$^{60}_{27}Co \longrightarrow {^{60}_{27}Co} + \gamma$$

Emission of gamma rays has no effect on mass and atomic numbers.

Top Tip

The sum of the atomic numbers and mass numbers must be equal for both sides of the equation.

Questions

1. Part of the decay sequence of uranium is described below.
   Uranium 238 decays by alpha-emission to form isotope X.
   Isotope X then decays, emitting Y to form protactinium 234. Identify X and Y.

Half-lives

What is a half-life?

The half-life of a radioisotope is the time taken for the activity or mass of a sample to halve. The symbol given to half-life is $t_{1/2}$. Even though the decay of individual nuclei within a radioisotope is random, the decay curves for all radioisotopes all follow the same (exponential) pattern.

Examples

1. The decay curve for isotope X is shown. Calculate the half-life for this isotope.
   200 counts/minute = 10 minutes
   100 counts/minute = 15 minutes
   50 counts/minute = 20 minutes

   It takes 5 minutes for the count to halve each time, so the half-life is 5 minutes.

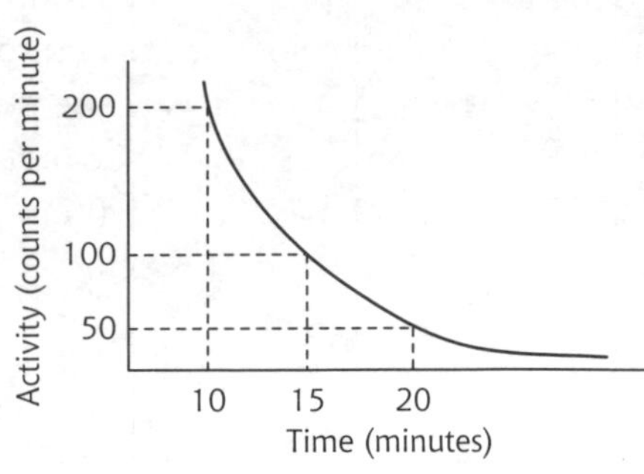

2. Bismuth-210 is a beta emitter decaying to form polonium-210. It takes 15 years for the bismuth's activity to drop to an eighth of its original value. What is the half-life of bismuth-210?
   Original activity = 1
   After 1 half-life = 1/2
   After 2 half-live = 1/4
   After 3 half-lives = 1/8
   Three half-lives = 15 years
   One half-life = $^{15}/_{3}$ = 5 years
3. Iodine-131 has a half-life of 8 days. Calculate the mass of iodine left in a 30 g sample after 24 days.
   Mass at beginning = 30 g
   After 8 days (1 half-life) = 15 g
   After 16 days (2 half-lives) = 7.5 g
   After 24 days (3 half-lives) = 3.75 g

Environmental effects on radioisotopes

Half-lives are not affected by temperature or pressure. Similarly they are not affected by chemical states, i.e. the half-life of polonium-218 is the same whether it exists as an atom of the element or as an ion in a compound. Different radioisotopes of the same element have different half-lives, e.g. lead-212 has a half-life of 10.6 hours whilst lead-214 has a half-life of 26.8 minutes.

The half-life and intensity of radiation are different. The half-life is for a particular isotope. The intensity is for the mass or concentration of the isotope present, i.e. 5 g of ^{14}C will have the same half-life as 40 g of ^{14}C but the intensity of radiation is greater for 40 g because there is more of the isotope present.

Questions

1. The half-life of lead–210 is 21 years. Calculate the mass left of a 40 g sample after 84 years.
2. Calculate the half-life of Rn–220 if it takes 220 s to decrease its radioactivity by $\frac{1}{16}$.

Radioisotopes

Artificial radioisotopes

Stable nuclei can be made unstable by bombarding them with particles such as protons, neutrons or alpha particles. These artificial radioisotopes usually have short half-lives.

Uses of radioisotopes

Use	Application and common isotopes
Medicine	^{60}Co is used to treat cancer whilst ^{133}I is used in the treatment of thyroid glands. Radioisotopes used in the body should have a short half-life and be beta or gamma sources because alpha particles can cause cellular damage.
Scientific research	Scientists can follow the path of a chemical reaction by making part of one of the reacting molecules from a radioactive isotope. The age of plants can be determined by carbon dating. The radioactive isotope 14 C is formed by the neutron bombardment of nitrogen in the atmosphere. $^{14}_{7}N + ^{1}_{0}n \longrightarrow ^{14}_{6}C + ^{1}_{1}p$ During photosynthesis, plants take in ^{14}C as part of their CO_2 supply. The level of ^{14}C in a plant remains constant during the life of the plant. When a plant dies there is no more uptake of CO_2 and the ^{14}C already in the plant decays. The proportion of ^{14}C changes, so the ages of plants can be determined by measuring this change. The half-life of ^{14}C is 5730 years.
Industry	Measuring the thickness of metals, detecting smoke and detecting cracks in pipelines are a few examples of the industrial use of radioisotopes. Radioisotopes used in industry should have long half lives, with the radiation type depending on the application, e.g. alpha sources measure the thickness of thin metal, whilst thicker pipelines require a gamma source.

Top Tip

In bombardment, the neutron is a reactant

Nuclear fission

Nuclear fission occurs when the nuclei of a heavy element are bombarded with neutrons. The nuclei split to form more, and lighter, nuclei, neutrons and energy, e.g.

$$^{235}_{92}U + ^{1}_{0}n \longrightarrow ^{94}_{38}Sr + ^{140}_{54}Xe + 2^{1}_{0}n$$

The energy given off heats water or gases, which turn turbines to produce electricity. The two neutrons released during the fission process bombard other nuclei. This results in the reaction becoming self-sustaining, i.e. chain reaction.

Nuclear fission versus fossil fuels

Fuel	Safety	Pollution	Sustainability of Resource
Fossil	More workers have died or had industry-related illnesses due to coal mining.	CO_2 (causing greenhouse effect)/ SO_2 (causing acid rain) produced by burning fossil fuels.	Supply of fossil fuels is limited. Fossil fuels produce a variety of products.
Nuclear	Has a high safety record – low possibility of a major accident.	Fall-out from nuclear accidents and problems in disposing of nuclear waste safely.	Large amounts of energy are produced from small quantities of uranium.

Nuclear fusion

Heavy nuclei are formed by the **fusing** of two lighter nuclei, producing energy but not the radiation emitted during nuclear fission. This process is not used to generate electricity, since a vast amount of energy is required to fuse the nuclei. Fusion takes place in the stars and this is how all naturally occurring elements, including those in our bodies, were formed.

$$^{2}_{1}H + ^{3}_{1}H \longrightarrow ^{4}_{2}He + ^{1}_{0}n$$

Answers

Energy Matters

Factors affecting rate 1

1. At the beginning, because the concentrations of reactants is greatest.
2. (*a*) Rate = 30 g/10 s = 3.0 g s^{-1}
   (*b*) between 21 and 22 s

Interpreting rate graphs

1. Activation energy.
2. Greater surface area.
3. (*a*)

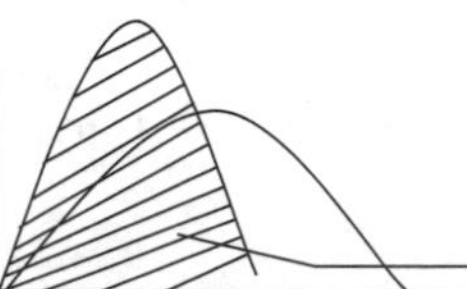

   (*b*) Average kinetic energy of particles decreases. Fewer particles have sufficient energy to overcome activation energy barrier.

PPAs

1. 1.8 s
2. The concentration was changed by diluting it with water but keeping the total volume the same, e.g. 25 cm^3 of KI, 20 cm^3 of KI and 5 cm^3 of H_2O.

Catalysts

1. A catalyst lowers the activation energy.
2. The active sites on a catalyst are blocked.
3. The reactants are in a different state from the catalyst.

Potential energy diagrams

1. (*a*) Reaction is exothermic.
2. (*a*) $E_a = 400$ kJ mol^{-1} $\Delta H = -200$ kJ mol^{-1}
   (*b*) $E_a = 600$ kJ mol^{-1} $\Delta H = +200$ kJ mol^{-1}
3.

Catalysed reaction

Calculating enthalpy change

1. (*a*) 668.8 kJ mol^{-1}
   (b) Loss of heat to surroundings or incomplete combustion
2. Mass of ethanol before and after combustion, volume of water and temperature of water before and after combustion
3. −736 kJ mol^{-1}
4. (*a*) $CH_4(g) + 2O_2(g) \longrightarrow CO_2(g) + 2H_2O(l)$
   (*b*) $Mg(OH)_2(s) \longrightarrow Mg^{2+}(aq) + 2OH^-(aq)$
   (*c*) $HNO_3(aq) + NaOH(aq) \longrightarrow NaNO_3(aq) + H_2O(l)$
5. Addition of CH_2 each time.
6. −43.47 kJ mol^{-1}
7. −58.6 kJ mol l^{-1}

The periodic table

1. 2204 kJ mol^{-1}
2. Increased influence of nucleus draws outer electrons in, resulting in atomic size becoming smaller.
3. Ionisation energies decrease. As you go down the group,
   (a) the outer electron is further away from the nucleus and therefore attached less strongly
   (b) the inner electrons will shiled the outer electrons from the positive nucleus.
   Therefore the energy required to release a mole of electrons decreases.
4. Electrons are being removed from a full shell of electrons.

Covalent bonding

1. & 2. (*a*) Polar covalent bond + polar molecule.
   (*b*) Pure covalent bond + non-polar molecule.
   (*c*) Polar covalent bond + non-polar molecule.

Structure

A metallic (lattice); B monatomic; C covalent molecules; D covalent network.
B buckminsterfullerene/allotropes of carbon

Properties

1. Delocalised electrons can move.
2. (*a*) pure covalent
   (*b*) polar covalent
3. Ions can move.
4. (*a*) Covalent network – covalent bonds throughout structure.
   (*b*) Covalent molecular – van der Waals forces between molecules.
5. Graphite has a covalent network structure and strong covalent bonds must be broken to melt graphite. Sulphur has weak van der Waals forces of attractions between molecules. These forces require a lot less energy to break.
6. Van der Waals.
7. Van der Waals forces of attractions increase. This is due to the atom size getting bigger (greater electron distortion).
8. More atoms, therefore more van der Waals forces.
9. NH_3 has hydrogen bonding, which is much stronger than the van der Waals forces of attraction which exist in PH_3.
10. Ethanol and hydrogen iodide.
11. Covalent molecular.
12. Abrasive.

13. (*a*) yes
    (*b*) no
    (*c*) yes
    (*d*) no
14. Hydrogen bonding forces water molecules into an open hexagonal shape – less dense.

Mole and reactants in excess

1. (*a*) 102 g
   (*b*) 132.1 g
2. (*a*) 0.5 moles
   (*b*) 2.5 moles
3. (*a*) 0.5 mol l^{-1}
   (*b*) 5 mol l^{-1}
4. 1.07 g
5. 16 g
6. 0.82 g
7. 0.1 mol l^{-1}
8. (*a*) Oxygen is in excess.
   (*b*) Sulphuric acid is in excess.
9. 2.76 g
10. 24.08 g
11. 0.6 g

The Avogadro constant

1. 3.01×10^{23} molecules
2. 1.2×10^{24} atoms
3. 6.02×10^{21} ionic units
4. 9.03×10^{22} atoms
5. 560 g
6. 306 g
7. 81.6 g of AlI_3

Reacting volumes

1. 44 litres
2. 60 cm^3 (30 cm^3 of O_2 left over and 30 cm^3 of CO_2 produced)

The World of Carbon

Fuels

1. Reforming
2. Naphtha
3. Small molecules

Alternative fuels

1. Advantage: burns completely; less CO produced/ less volatile than petrol.
   Disadvantage: less energy produced/extremely toxic/absorbs water so will corrode engine parts.
2. Means 'without oxygen'.
3. Made from sugar cane, which is a plant which can be grown year after year.

Hydrocarbons

1. (*a*) 2,2,3-trimethylbutane
   (*b*) 6,6-dimethyloct-2-ene
   (*c*) 7-methylnon-3-ene
2. (*a*)

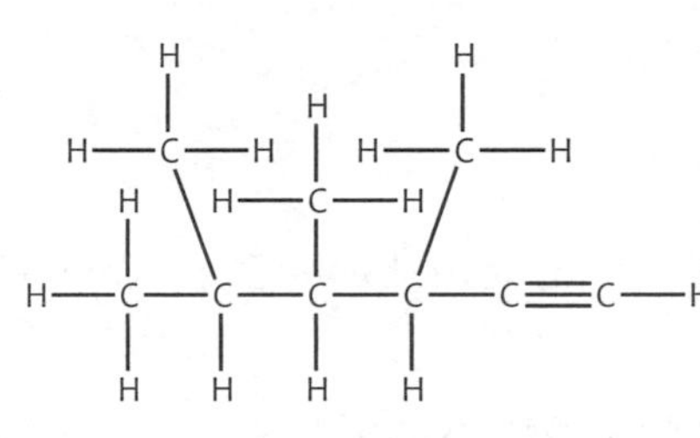

(*b*)

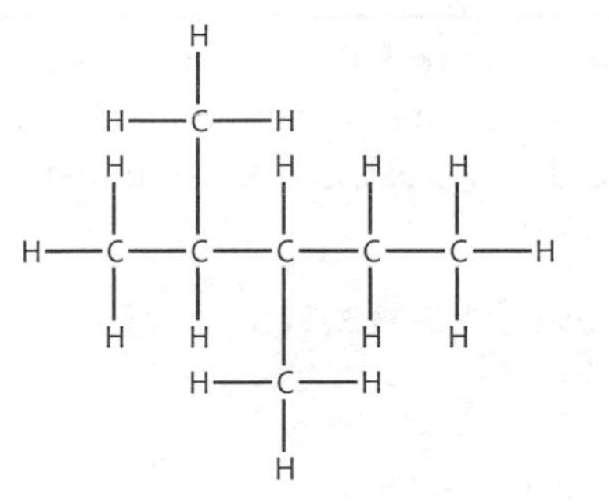

Alkanols

1. (*a*)

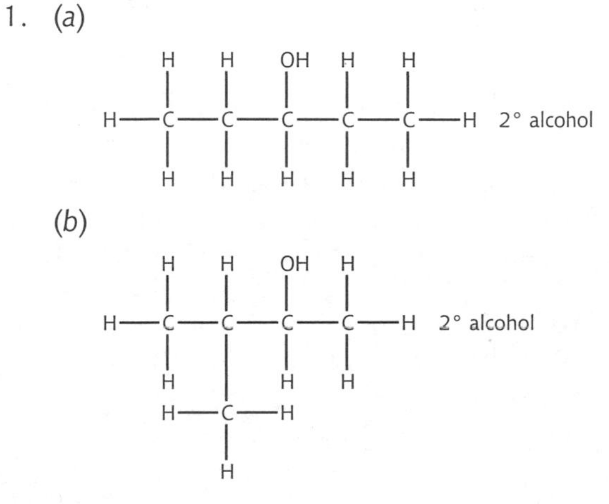

(*c*)

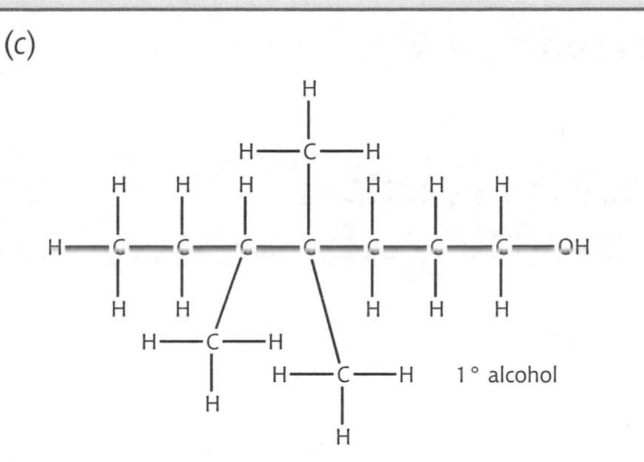

Aldehydes, ketones and carboxylic acids

1. (*a*)

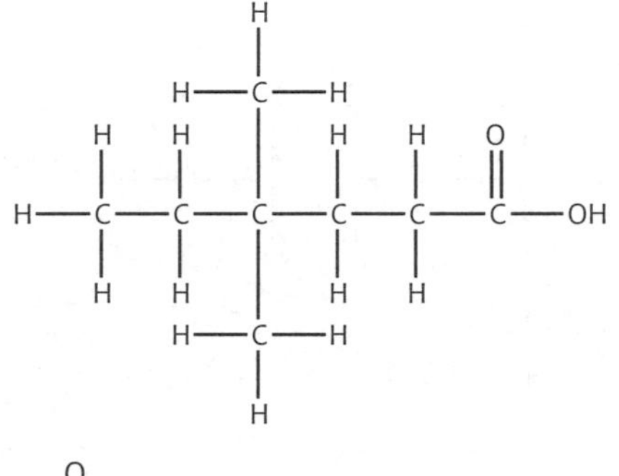

(*b*)

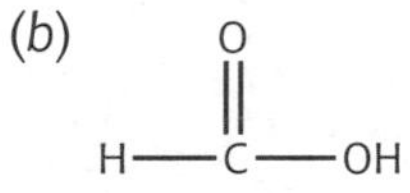

2. (*a*) butanoic acid
   (*b*) 3,4-dimethylpentanal
   (*c*) 4,6,7-trimethyloctanoic acid
3. (*a*) hexanal
   (*b*) 2-methylpentan-3-one
   (*c*) methanal
   (*d*) 2,3-dimethylbutanoic acid.

Esters

1. (*a*) hexyl butanoate

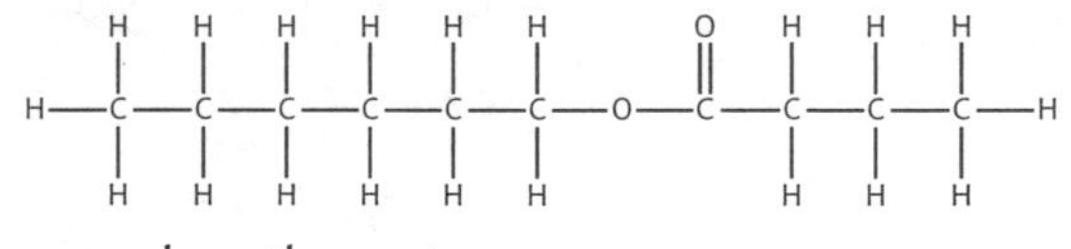

(*b*) propyl methanoate

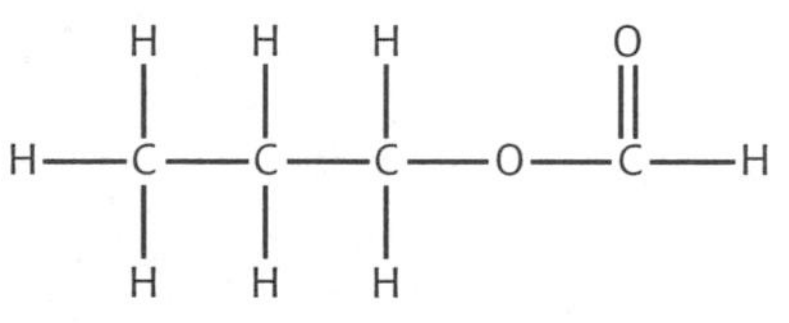

2. (*a*) ethanol and ethanoic acid

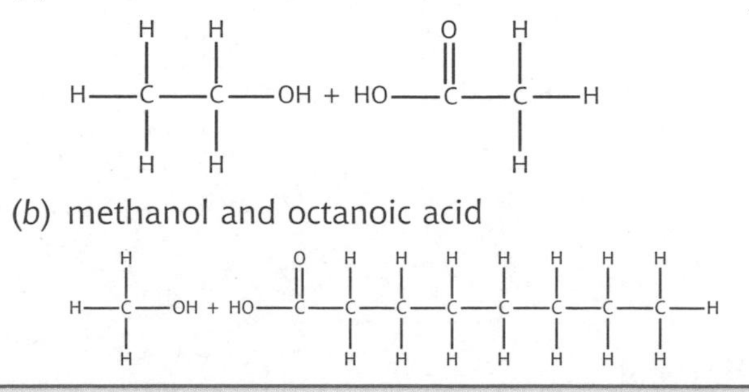

(*b*) methanol and octanoic acid

Percentage yield

1. 56.5%
2. 75%

Addition reactions

1. (*a*) ethanol

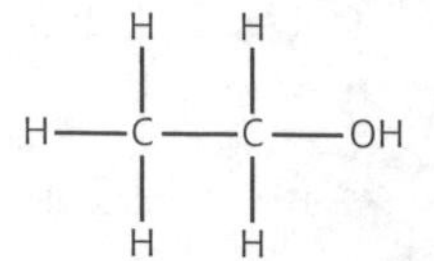

(*b*) 2-iodopentane and 3-iodopentane

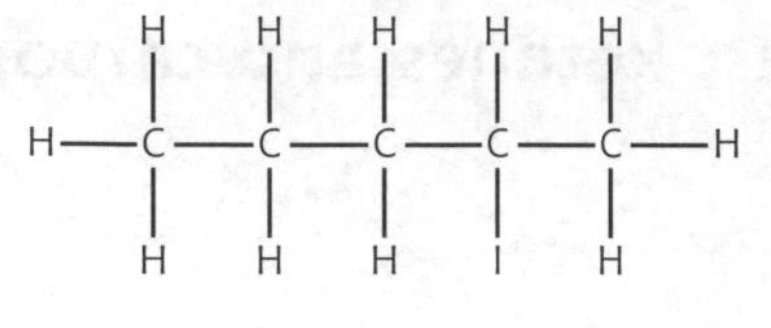

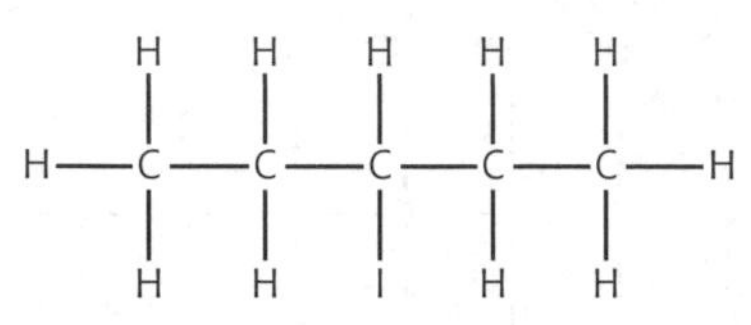

(*c*) butane

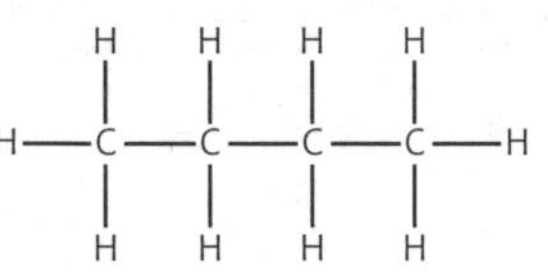

2. hydration
3. 1,1-dichloroethane and 1,2-dichloroethane
4. (*a*) 1
   (*b*) 1
   (*c*) 2
   (*d*) 2

Oxidation and reduction

1. (*a*) butanone

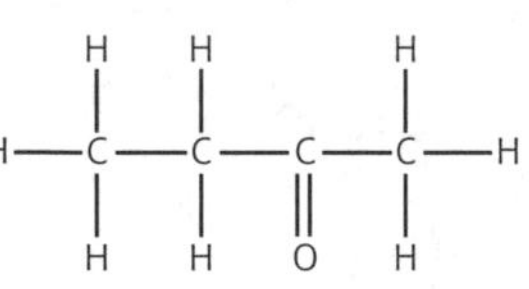

(*b*) methanal

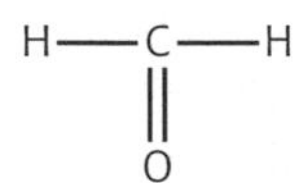

(*c*) propanone

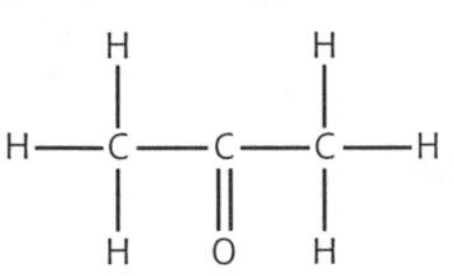

(*d*) no reaction

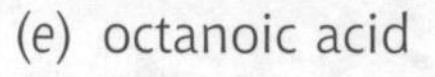

(*e*) octanoic acid

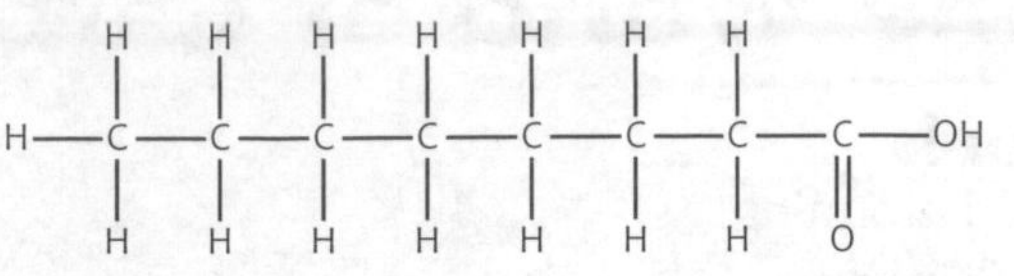

(*f*) no reaction

2. (*a*) methanol CH_3OH
   (*b*) hexan-3-ol $CH_3CH_2CH(OH)CH_2CH_2CH_3$
   (*c*) butan-2-ol $CH_3CH_2CH(OH)CH_3$
   (*d*) hexan-1-ol $CH_3CH_2CH_2CH_2CH_2CH_2OH$
   (*e*) pentanal $CH_3CH_2CH_2CH_2CHO$ and
   pentan-1-ol $CH_3CH_2CH_2CH_2CH_2OH$

Addition polymers

1. Cross-links between polymer chains cause a rigid 3-D shape.
2. Diacid and dialcohol.

Polyamides

1. Amino.
2.

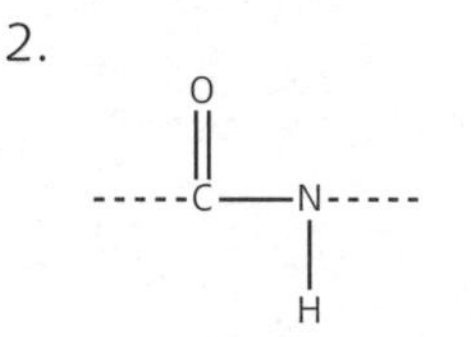

3. Hydrogen bonds form between polymer chains.
4. 4

Thermosetting plastics

1. The process has three steps:
   - *steam reforming* to produce synthesis gas
   - synthesis gas is then converted into methanol
   - methanol is then *oxidised* to form methanal.
2. Thermosetting.

Recent developments

1. (*a*) poly(ethenol)
   (*b*) kevlar
   (*c*) poly(vinylcarbazole)
   (*d*) biopol
   (*e*) poly(ethyne), poly(vinylcarbazole)

Fats and oils

1. Oil molecules can't pack close together, so have weak van der Waals' forces.
2. Glycerol or propane-1,2,3-triol.
3. 3.
4. Hardening.

Proteins

1. Amino acids.
2. Peptide link.
3. Fibrous.

Chemical Reactions

Types of process

1. Produce a variety of products, batch reactors can be used for slow reactions and are less expensive to build, reactants can be in a variety of states.
2. a (maintenance of plant).
3. b (sodium chloride).
4. Research and development, pilot study, scaling up, production and review.

Hess's law

1. The measurements for step 1: mass of KOH, volume of water, initial temperature of water and final temperature of solution. Step 2: volume of acid, volume of alkali, initial temperature of alkali, initial temperature of acid and final temperature.
2. Use an insulated cup (blown polystyrene) and lid.

Applying Hess's law

1. $-75\,kJ\,mol^{-1}$

The concept of dynamic equilibrium

1. No change in equilibrium position.
2. (*a*) Yield would drop as equilibrium position would shift to reactant side.
   (*b*) Yield would increase as equilibrium would shift to product side, because forward reaction is exothermic.

The pH scale and calculating the pH of acids and alkalis

1. Water contains only a very small amount of ions.
2. $0.001\ mol\ l^{-1}$
3. 11
4. 10^{-9}
5. 6.02×10^{21}

Strong and weak acids and bases

1. $HNO_3(aq) \longrightarrow H^+\ (aq) + NO_3^-(aq)$
   This equation shows that nitric acid fully dissociates.
2. Conductivity and pH of NH_4OH are low compared to KOH. This is due to partial ionisation of NH_3 which produces a small concentration of OH^- ions. Both alkalis will react with same volume and concentration of acid. This is due to the pool of OH^- ions which are produced by the full dissociation of KOH. NH_4OH molecules will dissociate to replace the OH^- ions that have reacted with the acid.
3. $CH_3COOH \rightleftharpoons CH_3COO^-(aq) + H^+(aq)$.
   This equation shows that ethanoic acid only partially dissociates.

The pH of salt solutions

1. (*a*) NH_4Cl = acid
   (*b*) CH_3COONa = alkaline
   (*c*) NaCl = neutral
2. $NH_4NO_3 \longrightarrow NH_4^+ + NO_3^-$
   $H_2O \rightleftharpoons OH^- + H^+$
   $NH_4^+ + OH^- \rightleftharpoons NH_3(g) + H_2O(l)$
   OH^- ions are removed from the water equilibrium. The water equilibrium moves to the right to replace the lost OH^- ions, producing an excess of H^+ ions and an acid salt solution.

Oxidising and reducing agents

1. $2Al(s) + 3Cu^{2+}(aq) \longrightarrow 2Al^{3+}(aq) + 3Cu(s)$.
2. Oxidising agent = Cu^{2+}
   Reducing agent = Al
3. $NO_3^- + 4H^+ + 3e^- \longrightarrow NO + 2H_20$
4. $FeO_4^{2-}(aq) + 8H^+ + 3e^- \longrightarrow Fe^{3+}(aq) + 4H_2O(l)$

Redox titrations

1. $4.8\,mol\ l^{-1}$

What is electrolysis?

1. 3.027 g of Cu
2. 0.253g of Al
3. 1329s or 22 minutes and 9 seconds
4. 317 minutes, 42 seconds (19 062 s)
5. 5.15 A
6. 0.291 A
7. $189{\cdot}6\ cm^3$

PPA quantitative electrolysis

1. To ensure a constant current was applied.

Types of radiation

1. X = $^{234}_{90}Th$
   Y = beta particle

Half-lives

1. 2.5 g
2. 55 s

My notes

03/280711

ISBN 978-1-84372-723-1

Published by
Leckie & Leckie Ltd
An imprint of HarperCollins*Publishers*
Westerhill Road, Bishopbriggs, Glasgow, G64 2QT
T: 0844 576 8126 F: 0844 576 8131
leckieandleckie@harpercollins.co.uk www.leckieandleckie.co.uk

Special thanks to
Integra Software Services Pvt. Ltd (creative packaging and illustration),
Helen Bleck (copy-edit and proofreading),
Peter Uprichard (content review).

A CIP Catalogue record for this book is available from the British Library.

Acknowledgements
Leckie & Leckie has made every effort to trace all copyright holders.
If any have been inadvertently overlooked, we will be pleased to make the necessary arrangements.